Seymour J Sharkey

Spasm in chronic nerve disease

being the Gulstonian lectures delivered at the Royal College of Physicians

of London

Seymour J Sharkey

Spasm in chronic nerve disease
being the Gulstonian lectures delivered at the Royal College of Physicians of London

ISBN/EAN: 9783337729349

Printed in Europe, USA, Canada, Australia, Japan

Cover: Foto ©berggeist007 / pixelio.de

More available books at **www.hansebooks.com**

SPASM

IN

CHRONIC NERVE DISEASE;

BEING

THE GULSTONIAN LECTURES

DELIVERED AT

THE ROYAL COLLEGE OF PHYSICIANS OF LONDON,

MARCH, 1886,

BY

SEYMOUR J. SHARKEY, M.A., M.B.Oxon., F.R.C.P.

ASSISTANT PHYSICIAN AND JOINT LECTURER ON PATHOLOGY AT ST. THOMAS'S
HOSPITAL.

LONDON

J. & A. CHURCHILL

11, NEW BURLINGTON STREET

1886

TO

EDMUND P. SHARKEY, M.D.,

THESE LECTURES

are dedicated

BY HIS SON

IN TOKEN OF

AFFECTION AND ADMIRATION.

SPASM IN CHRONIC NERVE DISEASE.

LECTURE I.

Mr. President and Gentlemen,—When I was informed that I had been nominated to fill the honorable position of Gulstonian Lecturer for 1886, I consulted the records of the College, in order that I might acquaint myself with the conditions attaching to the lectureship. I found that they were few in number, and that the main object which Dr. Gulston had in view was to secure, not only an annual dissertation on certain diseases, but also a practical demonstration of their pathological anatomy. It seemed to be incumbent, therefore, upon the lecturer, to choose a subject which he could illustrate, however imperfectly, from his own experience; and I selected that class of chronic diseases which have this, at least, in common, that they are associated with muscular spasm. To you, who know the wide range of knowledge and the ample experience which would be required to treat such a subject exhaustively, this proposal must appear pretentious to the last degree. I am fully aware, however, that the limited time at my disposal, and still more the limited knowledge I possess, make an exhaustive consideration of it impossible. I merely propose to take a general view of chronic nerve disease in its association with muscular spasm, and to consider how far our present knowledge of

1

pathology enables us to understand the phenomena which present themselves.*

By spasm is meant excessive muscular contraction; and, if we confine our attention to those muscles which are under the control of the will, excessive contraction is that which occurs in defiance of the will, or which, although commencing as a voluntary act, transgresses the limits which the will would impose upon it.

There is great variety in the spasmodic affections which have been observed in man, and to which descriptive names have been given. Thus, " tonic " spasm indicates persistent muscular contraction, a tetanic condition of a muscle, or of groups of muscles, in which the individual waves of contraction overtake each other, and are fused into one. " Clonic " spasm is a term applied to conditions in which the individual waves succeed each other less rapidly and often quite irregularly, giving rise to a great variety of intermittent spasms. The " co-ordinated spasms " form another class, in which groups of muscles are the seat of clonic contractions, which are co-ordinated in such a manner as to produce some regularly recurring though involuntary movement. It is unnecessary to proceed further in explanation of the many forms of pathological contraction of muscle which are embraced under the term spasm; they will be referred to again in connection with the diseases which they accompany.

In health, as a rule, contraction of striated muscles depends on impulses transmitted to them through the nervous system, and travelling down motor nerves; and the same may be said of that which is the product of disease ; so that the motor nervous mechanisms afford a suitable basis, in connection with which we may survey our present knowledge of the conditions which give rise to spasm. It is convenient, and I might almost venture to

* I am indebted to my friend, Dr. Theodore Acland, for many of the microscopic preparations from which drawings have been made, and to Mr. Shattock, Mr. Burgess, and Mr. Lapidge, for most of the drawings. Only a small number of the illustrations shown when the Lectures were delivered are reproduced here.

say scientific, to divide the motor mechanisms into a cerebral system and a spinal system ; the latter forming a groundwork on which the former is superposed, comparatively late in the history of the development of the individual, and probably of the species. It might be thought that, considering the importance of the cortical centres of the hemispheres, of the central ganglia, and of the cerebellum, the term cerebral systems would be more appropriate than cerebral system. But it is better, at any rate in considering spasmodic diseases, to embrace them all under one head. For our present knowledge of anatomy, physiology, and pathology, does not justify us in concluding that there is any efferent motor connection between the brain and the spinal cord except the pyramidal tract, direct and crossed, diseases of which give rise to chronic muscular spasm. Evidence in support of this view will appear in subsequent portions of these lectures.

The terms " cerebral system " and " spinal system " are intended to be anatomically exhaustive, and to comprehend all possible nerve-mechanisms by which spasm is produced. There should be no room for any further division of the subject. But our ignorance, especially of the finer molecular changes which accompany not only abnormal but normal neuro-muscular conditions, during action and inaction, necessitates some reference to those spasms which are called functional, and which will, we may hope, with advancing knowledge, become amenable to a more accurate classification. The divisions of the subject, then, as now proposed, are (1) spasm in connection with cerebral motor mechanisms ; (2) in connection with spinal mechanisms ; (3) functional spasm.

1. SPASM IN CONNECTION WITH CEREBRAL MOTOR MECHANISMS.

The nerve-fibres, which bring the voluntary muscles of the body under the control of the will, originate, so far as we know, in the central portions of the cortex. Experiments on living animals first yielded the clue to this discovery in man, and pathological evidence has abundantly corroborated the facts derived from this source. On an occasion like the present, it will not be thought a presumption if I refer to my own observations upon this point. In a paper published in the ' Lancet ' of September 29th, October 6th and 13th, 1883, I recorded six cases of cortical lesions all occupying the region in question, and all accompanied by motor paralysis, which occupied the face, arm, and leg, according to the localisation of the disease. It is unnecessary at present to enter into much detail on this subject. The two ascending convolutions which form the boundaries of the fissure of Rolando, the superior parietal lobule, the paracentral lobule, and probably some of the parts immediately adjacent to these, constitute the motor area of the cortex (Fig. 1). All we

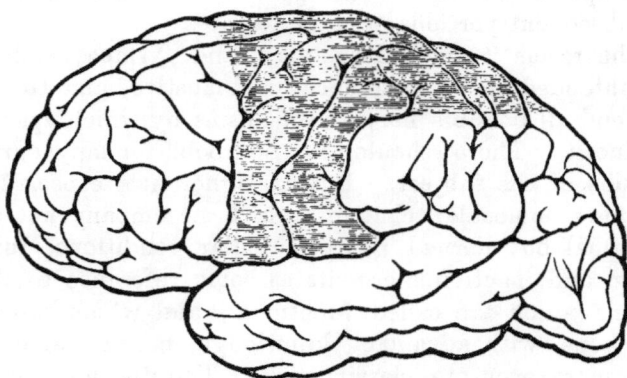

FIG. 1.—Diagram showing the position of the Motor Area of the Cortex on the External Surface of the Brain.

know about it, as regards voluntary muscular acts, is that the " way out " of the cortex for voluntary impulses passes through it. Some of the motor fibres are connected with processes of large ganglionic cells, which form a very striking feature of this region of the brain, and which resemble the cells of the anterior cornua of the spinal cord. The centres which are in connection with the leg occupy the upper part, those which are in connection with

the arm the middle, while the lower portions near the
fissure of Sylvius preside over the muscles of the face and
tongue.   During a part of the very long course which the
fibres coming from the motor area pursue, the bundles on
their way to the spinal centres for the face, arm, and leg,
remain distinct.   Thus, passing from the cortex through
the white substance of the hemispheres, they converge to
the internal capsule, and occupy the knee and anterior

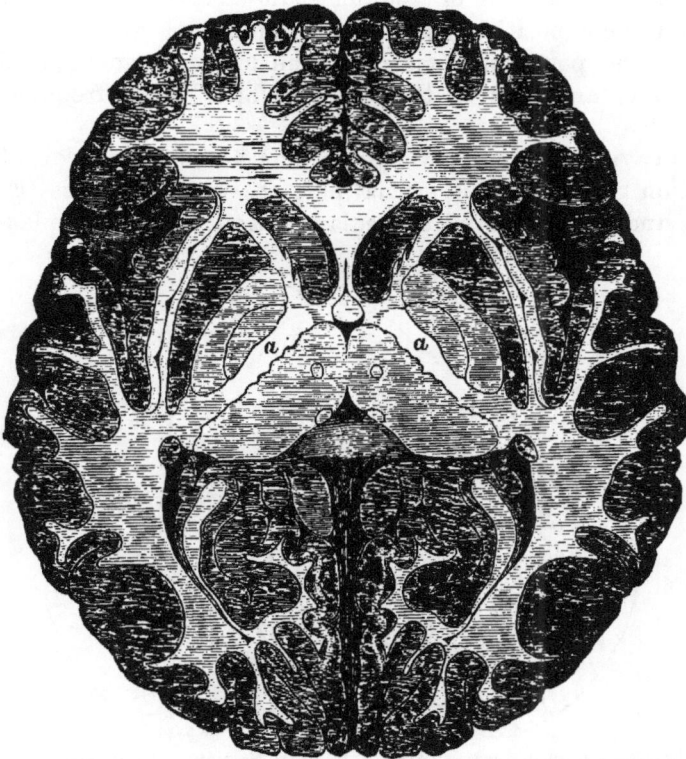

Fig. 2.—Horizontal Section of Brain (Flechsig).   Diagram showing the
position of the Motor Strands (*a a* unshaded) in the Internal Capsule.

two thirds of its posterior segment, which lie between the
optic thalamus and the lenticular nucleus (Fig. 2).   In
the capsule, the nerves for the face- and tongue-centres

occupy the knee, the fibres for the arm come next, and those for the leg lie posteriorly.   The knowledge of their relative positions in the internal capsule, which is a somewhat recent acquirement, is of great interest, because it probably affords at least part of the explanation of a long observed clinical, fact; namely, that, in hemiplegia in connection with disease of the central ganglia, the arm is usually more paralysed than the leg.   For hæmorrhages from the branches of the lenticulo-striate artery would, from their position, press more upon the anterior than upon the posterior division of the motor segment of the capsule, and this is the vessel which most frequently bleeds.

Leaving the internal capsule, the fibres under consideration pass to the superficial part (basis) of the crus (Fig. 3), and there occupy its median portion, and are so placed

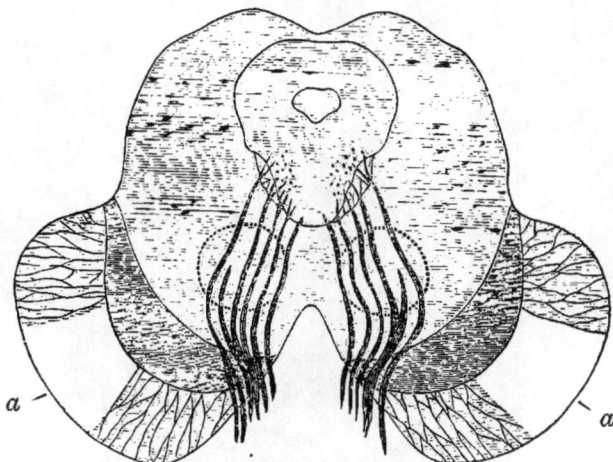

FIG. 3.—Diagram representing a Vertical Section through the Crura Cerebri : *a a*—unshaded arcæ—show the position of the Motor Fibres from the Motor Area of the Cortex.

that the fibres presiding over the face are situated internally, those over the leg externally, and those in connection with the arm centrally.

Arrived at the pyramids, in the medulla oblongata, the fibres going to the spinal motor centres for the body and limbs decussate, those for the face and tongue having already done so a little higher up. It has been estimated that about 91 to 97 per cent. decussate; but the number probably varies within wide limits, so much so, that there is good reason for thinking that in some cases complete decussation, in others none at all, takes place. Flechsig has also shown that, in the same spinal cord, the two sides may be unsymmetrical in this respect. Some observers suppose that decussation of fibres of the pyramidal tract occurs in the cord ; but nothing is known for certain with reference to this point. That part of the pyramidal tract which crosses at the decussation of the pyramids (the crossed pyramidal tract) occupies the posterior part of the opposite lateral column in the cord, being separated from the surface of the latter by the direct cerebellar tract. The fibres which do not decussate (direct pyramidal tract) pass down the cord on the same side, occupying a narrow strip along the margin of the anterior fissure (Fig. 4). The fibres of the pyramidal tract finally

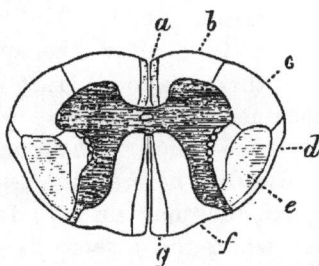

FIG. 4.—Diagram of a Section of the Spinal Cord in the Cervical Region (Flechsig). *a*, Direct Pyramidal Tract. *b c*, Antero-lateral Column. *d*, Direct Cerebellar Tract. *e*, Crossed Pyramidal Tract. *f*, Columns of Burdach. *g*, Columns of Goll.

form connections with the multipolar cells in the anterior cornua of the cord, either directly or through the intermediation of other nerve-cells.

It is a well-known fact that nerve-fibres in general depend for their integrity upon connections with ganglionic cells, and that, when these are severed, the fibres degenerate. Thus, the motor fibres of the cord depend upon the cells in the anterior cornua ; the sensory fibres upon cells in the ganglia on the posterior spinal roots ; while the healthy nutrition of the fibres of the pyramidal tract is dependent upon the cells in the motor convolutions. If their connection with the latter be anywhere interrupted, or if the cells themselves be destroyed, degeneration occurs in the whole tract below. When this has taken place, tonic spasm supervenes in the muscles, which are, in health, under the control of these fibres. But, between fully developed spasm and the earlier slighter indications of its presence there are many gradations. Taking, for instance, what is perhaps the commonest condition which gives rise to these events—destruction by a sudden hæmorrhage of the motor portion of the internal capsule—it is often some months before well-marked spasm is observed ; but, in the meantime, there are phenomena which foretell its coming. The most striking and frequent of these is the increased knee-jerk, which a tap on the patellar tendon elicits. The knee-jerk may be diminished for some time after the apoplectic attack ; but, if the fibres of the pyramidal tract have been interrupted, it soon becomes exaggerated. What are termed ankle-clonus and knee-clonus (phenomena which it is unnecessary to describe before an assembly like the present) soon appear, or may do so ; for they are not necessarily present when the disease is slight. Charcot reports a case where ankle-clonus appeared eight days after the hemiplegic attack, and late rigidity began within three weeks ; and another, in which it set in in one month, and late rigidity in two. Pitres gives cases in which clonus was observed in from eleven to fifteen hours. But, in any case, it may be said that increased tendon-reflexes and clonus precede rigidity ; and, in many cases of chronic disease of the lateral columns of the cord, they are to be found a very long time before rigidity, or the latter may not set in at all before death occurs. Sometimes one tap on the patellar tendon gives rise to a succession of knee-jerks, and like phenomena may appear in the other limb, or in muscles still further removed. Brissaud asserts that even the non-paralysed side is not healthy, but presents in a certain degree these phenomena, which he terms " latent contracture." To what cause are these increased tendon-reflexes to be referred ?

Muscular tone is that condition of tension in healthy muscle which is, no doubt, due to nerve-impulses proceeding from the cells in the anterior cornua of the spinal cord. A certain number of observers have come to the conclusion that the knee-jerk and other tendon-contractions are local phenomena, due to direct contraction of the muscle, which is temporarily put upon the stretch by tapping its tendon. Its increase is due to the increased tension which results from severance of the spinal centres from the cerebral motor centres, which exercise a controlling power over those below, this severance producing increased muscular irritability. But for the tendon phenomena to take place at all, the reflex arc producing tension must be intact.

Other authorities have convinced themselves, by equally careful observations, that the tendon phenomena are true reflex, neuro-muscular acts. The objection to this conclusion, which appears insuperable to their opponents, is that the latent period is much too short for a reflex act. But surely it is not legitimate to bind all reflex acts by the exact laws which appear to govern skin-reflexes. It is probable that conduction in nerves does not take place at precisely the same rate in every reflex arc in the body, and that even the time required to reflect the afferent impulse on to the efferent nerve is not uniformly the same. This will depend upon the purpose for which the muscular act is excited, and the speed with which it needs to follow upon the stimulus which evokes it. Now, there are, as has been proved, nerve-endings in muscles and tendons ;

and it is not improbable that the slight stretching of the
latter, which is produced by the action of antagonists,
gives rise, by means of a reflex act, to the steadying
action which is always aroused by the contraction of
opponents. This would require great sensitiveness on the
part of the reflex arc; for, considering the suddenness of
many muscular acts, the steadying contraction of the
muscles not directly producing the movement would require to be almost equally sudden. If the muscular act
which follows a tap on the patellar tendon be a purely
local phenomenon, due to "myotatic irritability," it is
difficult to explain those cases in which contraction of
muscles in the opposite leg, or even in the arm, occurs.
Gowers states that the latent interval for the knee-jerk is
$\frac{1}{25}$th to $\frac{1}{30}$th second, and that physiological data require
that it should be $\frac{1}{15}$th second if it be a true reflex. But
what will be said of the following equally legitimate argu-
ment? The interval for the knee-jerk is $\frac{1}{25}$th second.
Dr. de Watteville and Dr. Waller have found the interval
for the jaw-jerk to be $\frac{1}{50}$th second, which is only half the
interval required for the knee-jerk. Therefore, the phenomena are not produced in a similar manner. Dr. de
Watteville says that Dr. Waller has determined the
latency of the closure of the eyelids to the stimulus of a
strong light to be $\frac{1}{20}$th second, the shortest true reflex
known. This differs much less from the interval for the
knee-jerk ($\frac{1}{25}$th—$\frac{1}{30}$th second) than the latter differs from
the jaw-jerk ($\frac{1}{50}$th second).

We know too little about the various reflexes which
occur in the body to consider it as proved whether tendon
phenomena are reflex or not, but it is at any rate satis-
factory to feel that, for all practical purposes at the bedside, either theory is equally applicable. For the key-note
to both is integrity of the reflex arc, and changes in the
latter are necessarily accompanied by changes in the
tendon phenomena.

Let us assume, provisionally, that the knee-jerk and other
allied contractions are reflex. Why are they increased in

disease which interrupts the continuity of the pyramidal tract, or destroys the motor cells in which its fibres originate? Probably the reason is, not that the descending sclerosis produces irritability of the cells in the anterior cornu, but, as Hughlings Jackson says, that the suppression of the functions of the pyramidal tract gives rise to " permitted hyperphysiological activity," the spinal centres being " let go." This view is strongly supported by those cases of congenital absence of the whole or a part of the motor centres in the brain, which are accompanied by contracture of the limbs similar to that which occurs in lesions of the pyramidal tract resulting from disease ; for, in these cases, the pyramidal tract has been proved to be deficient. This was the condition found by me in a case which I shall relate presently. The pyramidal tract is not developed until very late, and is not even complete at birth ; and, in the cases of congenital disease referred to, it is not developed at all, or only in part. Even in disease which occurs in adult life and produces descending sclerosis, if the cord be examined after many years, the process of atrophy has long been completed, the nerve-fibres have disappeared, and nothing remains but a quiescent cicatrix. This was so in another case which I shall relate. In such conditions, no active process is going on to produce contracture ; there is merely destruction of the pyramidal strands. Again, the exhaustion following upon the motor discharge in epilepsy, may give rise to temporary increase in the knee-jerk, or to ankle-clonus ; and the same may be said of that suppression of function produced by the slighter degrees of chloroform narcosis. In short, the excessive tendon-reflexes and contractions which result from descending sclerosis are merely evidences of abolition of the function of the cerebral motor tract, which, in health, controls the action of the spinal ganglia.

Objections have been made to this view on the ground that sudden disease, such as cerebral hæmorrhage, which destroys the fibres of the internal capsule, does not give rise to the phenomena in question immediately, but only after the lapse of a certain time. The cases recorded by Pitres, in which ankle-clonus was observed to supervene in a few hours, prove that this delay is not an invariable rule; and, even if it were, it would not be unreasonable to suppose that cells, which have long been under the control of others, should require a certain time before they learn to exert a power to which they have hitherto been unaccustomed.

We may therefore look upon increased tendon-reflexes, clonus, and contracture as degrees of independent neuro-muscular action, originating in the spinal centres, which have run riot owing to the diminution or annihilation of the inhibitory functions of the pyramidal tract.

But, it may be further asked, does the appearance of one or all of these phenomena prove that there is gross and tangible disease of the cerebral motor cells or fibres? The answer is, decidedly not. In the 'Lancet' for November 7th, 1885, Dr. Angel Money records his experience of "Reflex Actions, Knee-jerks, and Muscular Irritability in Typhoid Fever, Phthisis, and other continuous Fevers;" and he shows that all the phenomena, except tonic spasms, are found in these conditions. But even contracture may occur in fevers. I have myself seen it in typhoid fever, and Dr. Murchison mentions its occurrence in typhus and relapsing fevers as well. Speaking of the former, he remarks : " Contraction and rigidity of certain muscles are observed more rarely, and only in severe cases. The fingers may be tightly clenched, or the forearm flexed, or, in rare cases, there is trismus or strabismus. In twelve cases, I have observed tonic spasms of many different muscles. Twice I have seen the legs and thighs so bent that the knees almost touched the chin." A man, suffering from extreme debility and anæmia, came lately to my out-patient room, and the house physician called my attention to the fact that he had not only very excessive knee-jerks, but as marked knee- and ankle-clonus as I ever saw. He was sent into the hospital; and after two days' rest in bed I was unable to elicit even the slightest clonus. Such cases as these, together with others in which recovery has occurred, justify the verdict which Dr. Donkin gives in ' Brain,' October, 1885. " It seems clear," he says, "at present that we must bear in mind that we have not done with spastic paraplegia by calling it lateral sclerosis, remembering the cases which recover, whether they are quite obviously functional or not, but must believe with Friedreich, Wilks, and others, that the symptoms may occur under various forms of disorder or disease of the nervous system."

My own impression is, that the condition which most frequently gives rise to these abnormal muscular conditions in connection with the spinal centres, is partial or complete suppression of nerve-impulses, passing from the cerebral motor centres down the pyramidal tract; and this may occur under a considerable variety of circumstances. But, accepting the fact that contracture of muscles, accompanying or succeeding a phase where tendon-reflexes are exaggerated, is most typically seen and is most characteristic in gross disease of the motor convolutions and pyramidal tract, let us consider some concrete instances of it. Two facts must be remembered, however, and these are, first, that when contracture is well developed, it may prevent us from eliciting ankle- and knee-clonus, and other tendon-reflexes; and, secondly, that contracture sometimes passes off, the muscles at the same time undergoing rapid atrophy, and the reflexes being abolished. This occurs when disease of the lateral column extends to the cells of the anterior column.

Case of Embolism of the Sylvian Artery, which resulted in Absorption of a large part of the Right Hemisphere of the Brain, and which was followed by Paralysis and Rigidity of Limbs.—The complete history of the patient's illness and of the post-mortem examination is recorded in the ' Transactions of the Royal Medical and Chirurgical Society' for 1884, and I shall not enter into much detail now. The case is an instance of extensive destruction of the cortex and central ganglia on one side, producing descending degeneration in the cord, together with the symptoms which are characteristic of that condition.

E. J—, æt. 34, was admitted into Guy's Hospital on March 8th, 1876, under the care of Dr. Pavy. She had never had any serious illness, and, though married ten years, had had no children. Suddenly she was attacked with giddiness, faintness, and sickness, and had an epileptiform seizure. After four days she recovered consciousness, and found she was paralysed on the left side, and had no sensation on that half of her body. On admission, she was found to have ordinary complete hemiplegia and hemianæsthesia on the left side. She had, in addition, a systolic murmur, audible all over the cardiac area. About four months after the commencement of her illness she left Guy's Hospital with the left arm almost useless, but with very fair power in the leg. She had completely recovered from the hemianæsthesia.

Seven years later, on May 7th, 1883, the patient, æt. 41, was admitted into St. Thomas's Hospital under the care of Dr. John Harley, with aortic disease and bronchitis. Dr. Hadden, the registrar, made the following note:— "The fingers of the left hand are partially flexed on the palm and decidedly stiff. The skin over the backs of the fingers is whiter and more glistening than it is on the right hand. The grasp on the left is feeble, and there is rigidity of the flexors of the forearm and arm. The tendon-reflexes are very brisk in the left, but are hardly obtainable in the right arm. The left leg is somewhat rigid, the patellar tendon-reflex brisk, and there is slight ankle-clonus. The tongue is protruded straight. Slight paresis is observed in the muscles on the left side of the face. There is no loss of sensation. On casual examination at the bedside, the left eye appears to be sound, and there is no hemianopia or colour-blindness. Tested with a watch, hearing is found to be sharp, and equal on the two sides." The patient died somewhat suddenly on May 17th.

The husband informed me that, from the period of her

attack, seven years before her death, she had been weak-minded and queer, doing unaccountable things, such as she had never done before, but that she remained a good-tempered, hardworking, and affectionate wife. On post-mortem examination, marked aortic disease (ulcerative endocarditis) was found, and there was great cardiac hypertrophy, the heart weighing 17¼ oz. The lungs were hyperæmic and œdematous at the base, and the lower portion of the right was semi-solid. The liver showed slight congestion in the centres of the lobules ("nutmeg"). The kidneys weighed 11 oz., the capsule was thick and adherent, the substance very firm, and the surface granular. The spleen was hard, and its capsule thick.

In the right hemisphere of the brain the parts supplied by the middle cerebral artery were extremely wasted, and many of the convolutions had entirely disappeared. The central ganglia, too, were so much affected, that the measurement from their intraventricular aspect to the outer surface of the convolutions of the island of Reil on the right, was not much more than one third of that on the left. Descending degeneration of the lateral column, and also of the column of Türck, could be seen in the spinal cord; and the alteration in both these strands of fibres was situated on the side opposite the diseased hemisphere.

This case, the brain of which is now shown, together with drawings of microscopic sections of the spinal cord, is of interest, not only because of the long time the patient lived after such extensive destruction of one hemisphere, but also because it is an instance of total decussation of the pyramidal tract. It will be seen on examination of the sections of the cord that the left half is smaller than the right, and that the white substance bordering on the anterior fissure, the column of Türck, is decidedly smaller on the left than on the right; that is to say, there is no direct pyramidal tract, but the whole of the fibres from the right hemisphere appear to have crossed, and to have gone, some into the left lateral column, some into the left column of Türck.

On microscopical examination, there is seen simply diminution in the number of white fibres, and increase in the density of the neuroglia in the affected lateral column. Apart from the irregularity which exists in the columns of Türck, there is very little structural difference to be made out in them. The softening, which produced hemiplegia and subsequent rigidity in this instance, involved the cortex, central ganglia, and internal capsule.

Lesions need not, however, be so extensive in order to give rise to paralysis with rigidity. Destruction of a portion of the motor convolutions, or of the subcortical fibres connected with them, or of that part of the internal capsule through which the motor strands pass, produces the same phenomena. Cases of softening of portions of the cerebral cortex of long standing do not often come under the observation of pathologists in the post-mortem rooms of general hospitals ; or, if they do, I have been singularly unfortunate in not meeting with them. A girl about eleven years of age, who was under my care as an out-patient at St. Thomas's Hospital, had had a very severe burn on the right side of the head when she was two years old. The scar occupied the greater part of the parietal region, and it was perfectly destitute of hair. I got no very clear details of the symptoms which succeeded this accident, except that she lost power all down the left side, a condition from which she had never recovered. Since then she had been subject to epileptic fits, in which the left side was convulsed ; but she did not lose consciousness. The arm, when I saw her, was rigidly contracted, the leg being much less affected, and the deep reflexes were excessive all down the left side. Now, although the actual state of things was not demonstrated by a post-mortem examination, it is pretty certain that the burn caused superficial disease of the parts situated immediately below and within the cranium ; and destruction of a portion of the motor cortex, with descending degeneration of the pyramidal tract, was the result.

Case of Perforating Tuberculosis of the Skull, with Cere-

bral Symptoms.—In 'Brain,' April, 1885, Dr. Edmunds describes a case, which affords a good instance of descending degeneration and rigidity from a limited cortical lesion. The patient was a boy of fourteen, of a delicate family. Eighteen months before admission into St. Thomas's Home, he had suffered from what was supposed to be tubercular peritonitis, and at the same time an abscess, accompanied by much headache, formed in the region of the left parietal bone. About six months later he complained occasionally of numbness in the right arm and hand, and two days before admission he had his first fit. He felt giddy, fell, became unconscious, foamed at the mouth, bit his tongue, and passed urine involuntarily. Both legs were convulsed, the right arm flexed, and the right hand clenched. He was admitted into St. Thomas's Home on September 3rd, 1884. An abscess was found situated over the left parietal bone. The grasp of the right hand was much weaker than that of the left, and there was very slightly deficient power in the right leg and right side of face. Sensation was everywhere normal, but double optic neuritis was present. I saw the patient with Dr. Edmunds at this time, and subsequently the abscess was laid freely open and a piece of dead bone removed, so that the dura mater was exposed and the pulsations of the brain could be seen. After this he had two fits, much like the one already mentioned. The discharge from the opening in the skull gradually diminished, and the scalp covered over the edges of the wound, so that when he left the hospital there was a defect in the skull and scalp, through which the dura mater could be seen. A silver plate was made to protect this. On February 4th, 1885, he left the Home, the chief symptom of disease which he presented being partial paralysis of the right arm and hand ; when tested with the dynamometer the right hand only marked 20, the left 70.

In the ' Lancet' for September 27th, 1884, my colleague, Dr. Reid, published his " Observations on the Relation of the Principal Fissures and Convolutions of the Cerebrum

to the Outer Surface of the Scalp." Before the patient
left the Home, Dr. Edmunds got Dr. Reid's assistance in
determining the position of the lesion by surface markings.
The boy's head was shaved, and the fissures marked upon
the smooth surface, and the opening in the skull was found
to be half an inch in front of the middle of the fissure of
Rolando.   But, as it was known that the disease extended

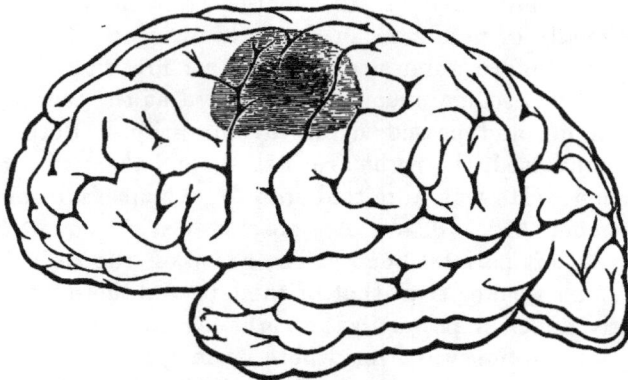

FIG. 5.—Diagram showing the position of the Cortical Lesion.

backwards along the dura mater for a short distance behind
the opening, the conclusion drawn was that the middle of
the ascending frontal and ascending parietal convolutions
was the region affected.   The account I am about to give
of the post-mortem examination shows how accurate this
localisation was.   The patient returned to the Home on
September 11th, 1885, looking very ill and emaciated, and
with a sinus on the right side and back of the neck, connected with an evident angular curvature of the spine.
His right arm was flexed at the elbow and could not be
completely extended.   He died of exhaustion on November
12th, 1885, the end being accelerated by diarrhœa, which
was scarcely controllable.   At the post-mortem examination, relics of past peritonitis were found and old and recent
tubercular affection of the lungs.   There was a chronic
abscess in relation with caries of the lower cervical ver-

tebræ, the disease which had produced the angular curvature. There was chronic thickening of the membranes of the cord in this region, but no evident disease of the cord itself. Dr. Edmunds was good enough to hand over to me the brain and spinal cord for careful examination. The following is an account of the condition of things which I found after hardening the specimens in spirits.

The dura mater, which was thickened and adherent over a portion of the left parietal convolutions, was gently peeled off, and that part of the cortex which was so far invaded by disease that it could not be separated from the dura mater, but was torn during its removal, is marked in Fig. 5. The lighter shading represents very slight superficial abrasion, and the darker central part shows where the disease had sunk deeper into the grey matter. The area marked in the figure measured an inch and a half vertically, and the same horizontally. The position which it occupied was the upper half of the middle third of the ascending frontal and parietal convolutions, together with a small portion of the posterior extremity of the middle and of the superior frontal convolutions. On making a horizontal section through the basal ganglia and the internal capsule, a slight discolouration could be seen in the latter just behind its knee. On making a vertical section through the ascending frontal convolution and traversing the centre of the diseased area, the destruction of cortex was seen to be quite superficial, and no degeneration was made out in the subcortical white matter. The cord was examined high up in the cervical region, above the seat of chronic meningitis which resulted from the disease of the cervical vertebræ. The sections show that the right half is much smaller than the left, the diminution in size being apparent, not only in the lateral column, but also in the column of Türck. Under the microscope, the connective tissue in the right lateral column is seen to be denser than that in the left, and the nerve-fibres appear to be here and there less numerous.

Hospitals where old people congregate afford the best

opportunities for post-mortem examination of persons whose brain is the seat of softening originating long before death. Charcot has given his experience at the Salpêtrière, and he says (' Lectures on the Localisation of Cerebral and Spinal Diseases,' translated by Walter Baugh Hadden,M.D., 1883, New Sydenham Society, p. 141) : " In these cases, the lesion presented itself under the form of yellow softening (*plaques jaunes*), more or less extensive in area, involving, to a variable depth, the subjacent white matter, and occupying the most diverse regions of the surface of the hemisphere. It is expressly mentioned in all the observations that the softening had not affected the central masses : optic thalami, caudate and lenticular nuclei, and internal capsule. My observations may be divided into two groups.

" The first includes the cases in which permanent hemiplegia had not existed during life, and in which secondary degeneration was found at the necropsy to be absent. In all, the convolutions supplied by the Sylvian artery, and particularly the ascending frontal and parietal, remained intact. The yellow softening was situated in one of the following regions, namely, some part of the sphenoidal lobes, the lobulus quadratus, the cuneus, one or both occipital lobes in their entirety, some part of the anterior two thirds of the frontal lobes.

" In all cases of the second group, there had been, on the other hand, permanent hemiplegia and well-marked secondary sclerosis. The feature common to these cases is that the lesions invariably involved, to a greater or less extent, one or other of the ascending frontal and parietal convolutions, principally in their upper half, and often both at the same time. In addition, the regions nearest to the frontal and parietal convolutions were very frequently involved."

In most instances, destruction of a portion of the pyramidal tract on one side gives rise to paralysis and spasm on the opposite side of the body only ; but occasionally spasm of the muscles on the same side as the lesion follows

later in the history of the case. This is probably due to
sclerosis appearing in the opposite lateral column, as Pitres
has shows that it often does ; and Mr. Sherrington has
shown that it occurs in dogs which have had one pyramidal
tract experimentally injured. But, although this is probably the cause of most cases of bilateral spasm from unilateral lesion, the possibility of the presence of a bilateral
lesion must be borne in mind. In the ' Revue de Médecine '
for May, 1883, Émile Demange records a case of the
kind where a bilateral lesion was found in the brain post
mortem.

Leaving affections of the cortical and subcortical portions
of the pyramidal tract, let us pass to a consideration of the
symptoms which accompany diseases of it as it traverses
the basal ganglia, crura cerebri, pons Varolii, and medulla
oblongata.

It is needless, before an audience like the present, to
recount cases of hemiplegia followed by rigidity, which
result from destruction of the internal capsule by hæmorrhage or softening. Such examples of disease are of
common occurrence and well known to all. I shall rather
give an account of a series of cases which are less frequently met with, and which made a considerable impression upon me from two points of view :

1. Because they seem to throw light on the rigidity
and tetanic seizures which are observed in tumours of the
cerebellum.

2. Because I failed to recognise the position of the
disease in the brain during life, though a reconsideration
of the symptoms leads me to think that others might have
succeeded.

*Case in which a Tubercular Mass occupied both Optic
Thalami, producing Paralysis, Tremors, and Rigidity of
the Limbs.*—F. P—, æt. 4, was admitted into St. Thomas's
Hospital under Mr. Sydney Jones, in August, 1878, and
on the 22nd of the following October excision of the left
astragalus was performed. The child did very well, and
was bright and playful until January 28th, 1879, when

tremors of the left arm and leg were noticed. They were rhythmical to-and-fro movements, of short excursion, occurring only when the limbs were used voluntarily. There was very little loss of power, no affection of speech, deglutition, or respiration, and the tongue, face, and eyes all moved normally.

On March 5th the child was still bright and lively, though pale; and it was noticed that the tremors were more marked on some days than on others. The left side of the face and the right hand had also begun to show similar tremors when used. The patient was sent out of the hospital, but was readmitted under Mr. Sydney Jones on July 8th, 1879, because the ankle and general condition were said to be worse. He had gone gradually down-hill, had become very drowsy, passed his evacuations in bed, and was more tremulous. On readmission, his face was pale, and a large dilated vein coursed transversely across the right upper eyelid. He was always asleep unless roused, and would even fall asleep while eating ravenously. His appetite was good, but he never asked for food. He complained of no pain, but appeared to have general hyperæsthesia of the skin. He was very irritable and peevish, crying out loudly when disturbed or contradicted. There were now rhythmical tremors of the right upper extremity, which were excited by voluntary efforts, and resembled those which used to affect the left side; the latter had ceased and given place to rigid contraction. The elbow was flexed and the hand clenched so tightly that the finger-nails had produced ulcers in the palm. The hand was pronated and the rigidity was constant. The mouth appeared natural when at rest, but when the child laughed or cried it was drawn up to the right. There was no squint; the pupils were large and acted to light; the tongue was protruded straight but was tremulous, and speech was low. There was no deafness, but the ophthalmoscope revealed double optic neuritis.

There was no vomiting or dysphagia; the bowels were very confined, and the evacuations passed in bed. The pulse was 114, respirations 24, and the temperature varied between 99·4° and 101° Fahr. The viscera appeared to be healthy. Until October 8th the symptoms already described became worse. The right hand, arm, and leg trembled violently when used, and the left leg became perfectly rigid. The patient was very drowsy and seemed to suffer from some headache ; but he was never sick, and his appetite was good. His head was slightly retracted.

On October 8th he seemed much worse ; he vomited three times, took his food badly, and the temperature rose to 103·2° Fahr. Subsequently he became more and more drowsy, his right eyelid drooped, the pupils became unequal, and repeated attacks of general rigidity (sometimes accompanied with foaming at the mouth) supervened, and he died on October 15th.

On post-mortem examination I found a recent, acute, basilar meningitis. Besides this affection, which was, no doubt, the immediate cause of death, a large caseous tubercular mass was seen involving both optic thalami and almost completely obliterating the third ventricle. The right optic thalamus was the more completely disorganised of the two. The specimen which I now show was hardened in alcohol ; and, on section, it was then seen that on the left side the internal capsule, though pressed upon by the new growth, was clearly defined and looked fairly natural, while that on the right was unrecognisable. The tubercular mass was confined to [the optic thalami, and did not involve either nucleus of the corpus striatum. A very small caseous tubercle was also found on the internal aspect of the left occipital lobe, but the rest of the organs were healthy.

It is well known that diseases of the optic thalami and corpora striata are not accompanied by any such symptoms as were present in this case, unless the internal capsule is involved, and the paralysis, tremors, and rigidity of limbs must be attributed to the invasion of this strand of fibres proceeding from the motor convolutions ; while the convulsions of the last week were the result of the recent

meningitis. This case has been already recorded by Dr.
Bristowe, in ' Brain,' vol. vi, p. 167.

*Case of Large Caseous Tubercle of the Pons Varolii, pro-
ducing persistent Rigidity of the Neck, Back, Arms, and
Legs.*—C. C—, æt. 2, was admitted into St. Thomas's
Hospital under the care of Dr. Ord, on October 4th, and
died on November 7th, 1881. The mother was attending
Brompton Hospital at the time, and she had lost one of
her three children from consumption.

The patient was healthy until three months ago, when
he began to have diarrhœa and to lose flesh, but had no
vomiting. A week before he came in he had two convulsive fits. He had had pain in the head and had been
very fretful and drowsy.

On admission, the child lay in a dull and listless condition, taking but little notice of anything, and only crying
when disturbed. The head was drawn strongly backwards
and kept stiffly in that position. The back was likewise
rigid. The arms were flexed and the legs and feet extended, all being quite rigid. Patellar reflex was very
brisk. There was no loss of sensation ; deglutition appeared
to be imperfectly performed. The pupils were dilated,
but equal ; and Mr. Nettleship pronounced the fundus of
both eyes to be natural. Pulse 120. Loud rhonchi were
heard over both lungs. The urine was free from albumen.

During the rest of his life he became more and more
drowsy and had varying degress of hyperæsthesia, but
never loss of sensation. Rigidity of neck, trunk, and
limbs continued without intermission, though it varied con-
siderably in degree. Slight left internal strabismus and
partial paralysis of the right facial nerve came on ; and
deglutition was more and more difficult, the food running
out at the corners of the mouth. The fundus of the eyes
was repeatedly examined, even as late as November 6th,
but no changes were found. The temperature, with very
rare exceptions, was normal.

On post-mortem examination, I found the cerebro-spinal
fluid much increased and the lateral ventricles moderately

distended with serum. There was slight basal meningitis, but the hemispheres were free from tubercular disease. In the cerebellum and pons were large greenish tubercular masses of rather irregular outline, that in the latter being in the position marked in the figure. It occupied almost the entire extent of that division of the brain (Fig. 6),

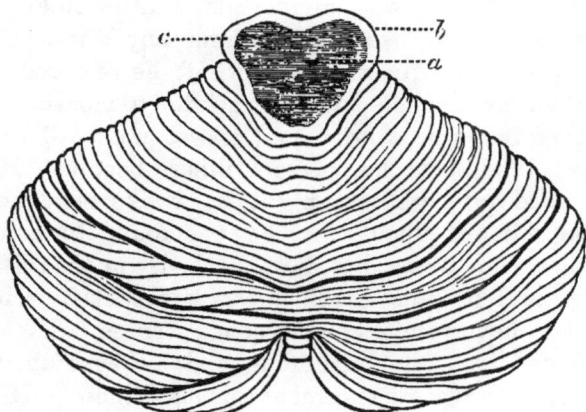

Fig. 6.—Diagram showing the position of a Tubercular Mass (*a*) as seen in a vertical section through the Pons (*b*). The Rim (*c*) represents the remaining Nerve Substance.

having only a thin sheet of white matter around it. Longitudinally, it extended the whole length of the pons, but seemed to stop short of the medulla oblongata. Unfortunately, no microscopic examination of the spinal cord was made. The lungs were thickly studded with miliary tubercles, but the other organs appeared to be healthy. The masses in the cerebellum were all situated superficially in the cortex and did not extend deeply into the substance of the organ.

In this instance, again, the muscular rigidity must be referred to pressure exerted by the tumour upon the fibres of the pyramidal tracts in the pons, for tumours situated in the superficial parts of the cerebellar cortex are of pretty frequent occurrence, and are not found to give rise to any such symptoms. I shall presently relate a case in proof of this assertion.

Case in which a Tubercular Mass occupied the Aqueduct of Sylvius and produced Tremors, Paralysis, and Spasmodic Contraction of Muscles.—E. B—, æt. 7, was admitted into St. Thomas's Hospital under the care of Dr. Bristowe, on August 2nd, 1882, and died on November 30th in the same year.

His family history was unimportant, and the child himself had never suffered from fits or from any serious illness, except from a blow on the head which he received at the age of four and which laid him up for a month. From this he perfectly recovered, and it was not until the end of May, 1882, that his present illness began. Having gone to bed quite well one night, he awoke next morning throwing himself about, and he was found to have lost power in all his limbs, but most markedly in the right arm and leg, and he was unable to walk properly. He was said not to have vomited or lost consciousness. His speech was affected from the beginning, and his eyes about one week after the commencement of his illness. He was becoming weaker and thinner every day, and the tremors in his limbs, which were noticed at first, were becoming steadily worse. His bowels were confined, and he had lost control over his bladder ; he was very drowsy, but his memory was good. He had all along been free from headache, vomiting, and convulsions.

On admission, the child was found to be rather thin and drowsy. He lay apathetically in bed, but could be easily roused if spoken to. He had weakness in the arms and legs, and, when he walked, his legs seemed to drag. The most striking peculiarity in his condition was, however, the tremor with which all his movements were accompanied. The head, neck, jaws, trunk, arms, and legs, were all similarly affected. When he used his hands and arms the tremors in them were marked, of rather short excursion, sometimes in the direction of movement, and sometimes at right angles to it. The jaws trembled when he used them, and speech was slow and drawling. The tongue was protruded straight, and, when he smiled,

the mouth was drawn somewhat to the right; hearing
was normal. All the muscles of the eyeballs were more
or less weakened. There was ptosis on both sides, and
the internal rectus of the right eye was markedly affected.
The superior and inferior recti on the right were likewise
weak, as were the muscles of the left eye, though not so
weak as those on the opposite side. Both pupils acted
well to light and to accommodation, the right being a
little larger than the left; no anæsthesia was present.
The patellar and plantar reflexes were normal, and there

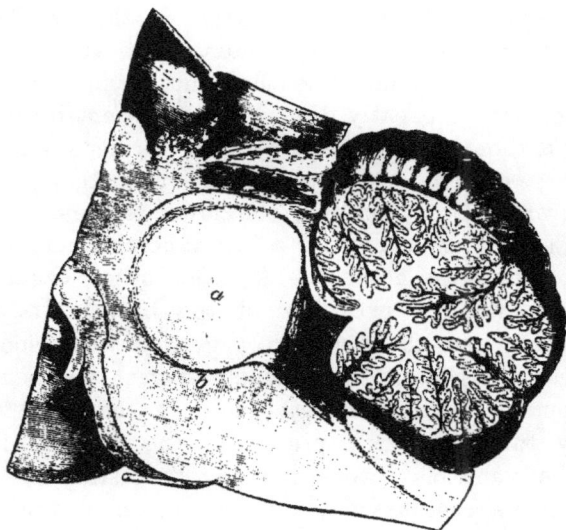

FIG. 7.—Drawing of a large Tubercular Mass (*a*) occupying the Aqueduct
of Sylvius, and by its pressure producing a Cup-like depression in
the Pons (*b*). The Section is Vertical and Longitudinal.

was no ankle-clonus. The temperature was natural, and
all the organs, except the brain, appeared to be healthy.
During the rest of his life the tremors gradually increased, paralysis became more marked, and attacks of
spasmodic contraction of the muscles of the limbs and
trunk occurred from time to time. On some occasions,
the legs were noticed to remain for a while rigidly

extended.   The patellar and plantar reflexes became
exaggerated.   The patient grew more and more drowsy,
and passed his evacuations in bed.   The temperature
presented only occasional slight elevations until November
26th, four days before death, when it rose to 101·2° Fahr. ;
and it subsequently reached 103·8° Fahr. on the 29th, the
day before death.

Throughout the patient's illness there was a singular
absence of headache, vomiting, and optic neuritis.   Though
frequently examined, the fundus of the eye presented no
abnormal appearance until November 26th, when there
was thought to be neuritis.   After death, Dr. Edmunds
found distinct microscopic evidences of this condition.
Another curious point was that the tremors were now
and then either greatly diminished or absent for a day or
two, and then returned with all their former intensity.

Necropsy.—The body was much emaciated, and rigor
mortis was present only in the lower extremities.   The
spleen and lungs contained a few tubercles, and those in
the spleen were caseous.   All the other organs were
healthy, with the exception of the brain.   Its surface
was sticky, and the veins were gorged with blood.   At
the base, in the interpeduncular space, the membranes
were opaque, and there was fluid in the optic sheath.
Miliary tubercles were present in the Sylvian fissures,
and the ventricles were moderately distended.   There
was a large caseous mass in the region of the corpora
quadrigemina, which was round in shape, and as large as
a medium-sized marble, and which is well displayed in the
preparation before you (Fig. 7).   It lay as if wedged in
the aqueduct of Sylvius, which was enormously distended.
On the upper surface of the tumour were the thinned and
flattened corpora quadrigemina, while its lower rounded
margin rested upon the upper surface of the pons and
crura cerebri.   Here it occupied a cup-like cavity, which
its downward pressure had produced.   No other important
structures were injured by the tumour.

In this case the usual phenomena of pressure upon the

fibres of the pyramidal tract are again well represented. Paralysis and tremors occurred first; then attacks of rigidity. Had the child lived longer, and the growth of the tubercular mass progressed, the pressure exerted by it upon the pons and crura cerebri would, no doubt, have produced the final stage of persistent rigidity. The recent tubercular meningitis cannot be said to have produced the tetanic seizures, as the latter occurred months before the patient's death. This case and the next which I shall relate have already been published by Dr. Bristowe in 'Brain,' vol. vi, p. 167.

Case of Tubercular Mass occupying the Left Lobe of the Cerebellum, and pressing upon and producing softening of the Pons and Medulla; Persistent Rigidity of Limbs.— A. G. J.—, æt. 4, was admitted into St. Thomas's Hospital, under the care of Dr. Bristowe, on December 30th, 1878, and died on August 6th, 1879. He had been a healthy child until three months previously, when he was ill with measles and whooping-cough. He began to have fits, which occurred about once a day at first, and then twice a day. He had also been suffering from severe pain in his head, occasional vomiting, and for about two months his mother had noticed that he squinted. The father was said to be asthmatic and the mother had a weak chest, but their two other children were healthy.

On admission, the patient seemed to be a sensible child, but quite blind. The pupils were widely and equally dilated, and contracted slightly to light. When lying in bed he could move his arms and legs, but the former were tremulous and ataxic. When placed upon the ground the patient laughed hysterically, but could not stand, though he could move his legs forwards when supported. The limbs appeared to be equally affected on the two sides. The mouth dropped slightly on the right side, and the tongue was protruded to the right.

The pulse was 140, and regular; respirations were easy. The patient passed all his evacuations in bed. There was well-marked optic neuritis in both eyes, the outline of the

discs being irregular and blurred, the veins large, the
arteries small, and the whole presenting a streaky appearance. Between the time of admission and his death the
patient had many fits, in which he was unconscious, the
legs remaining rigidly extended, and the arms being
flexed, and then extended several times in succession. A
certain amount of left facial paralysis developed, so that
the patient could not close the eye of that side, and the
left external rectus was partially paralysed. His arms
became more tremulous, the right being more affected
than the left ; his speech was very drawling. He remained
for long fairly intelligent, had no deafness, and was not
much troubled with headache or vomiting. From February
onwards the arms and legs became persistently rigid,
although the degree of rigidity varied. Finally, the child
became emaciated to the last degree, and lay perfectly
still, taking food well when it was offered him, but showing
very few signs of life. He died on August 6th.

Post-mortem examination.—There was little evidence of
important pathological changes in any of the viscera
except the brain. The head was very large, all the
sutures were separated, and the bones were moveable.
There was no subarachnoid fluid, but the ventricles were
greatly distended, and the substance of the hemispheres
reduced to a thin, soft sheet of nerve-substance. The left
lobe of the cerebellum was entirely occupied by a firm
tumour, of the size of a Tangerine orange, yellowish green in
colour, semi-transparent, and consisting of closely placed
concentric layers. The relics of cerebellar tissue remaining on its surface were soft, as was the neighbouring
part of the posterior cerebral lobe. The tumour was
adherent to the tentorium cerebelli, and the pons and
medulla were pressed upon by it, and somewhat softened.
The right lobe of the cerebellum was healthy. Another
small tubercular mass was found in the grey matter on the
orbital surface of the right frontal lobe.

In this case, again, persistent rigidity was the result of
the pressure which the tumour exercised on the medulla

oblongata and the pyramidal tracts which run in it. For it will be shown presently that without such pressure, tumours of the cerebellum do not produce this symptom, and that distension of the ventricles with fluid is not sufficient to explain it.

These four cases appear to me to be of great interest as representing diseases affecting a certain portion of the brain, and capable being localised during life. The region referred to extends from the basal ganglia to the medulla oblongata, commencing at that spot where the pyramidal tracts begin to converge as they pass between the central ganglia, and occupying further on the pons and medulla oblongata, where the pyramidal tracts lie in close contact. If the motor tracts be affected simultaneously, or quickly one after the other, in the way about to be mentioned, the disease is likely to be situated between the basal ganglia above and the medulla below, and is, in all probability, a tumour. The earliest symptoms referable to the pyramidal tract are tremors ; then come paralysis and rigid contraction, which is finally permanent ; and besides these, there may occur from time to time attacks of tetanic spasms, affecting most of the muscles of the body. The phenomena of chronic spasm are due to gradual pressure on the motor tracts, proceeding from the central convolutions of the brain ; and the fact that both sides of the body are attacked indicates that the disease is situated at a part where the pyramidal tracts are in close proximity to each other. If the case be seen early, probably all these symptoms, tremors, paralysis, and rigid contraction, will occur before the end ; but if the patient does not come under observation until late, only some of them may be observed. Thus, in the case where the optic thalami were gradually invaded, and in the case of cerebellar tumour pressing from above on the medulla, all occurred ; whereas, in the case of the large tubercular mass almost completely occupying the pons, only the later stage of persistent rigidity was seen. In determining the exact seat of the tumour, whether in the optic thalami, corpora striata, pons, or cerebellum, we must consider the other symptoms referable to the special nerves involved, which arise at different levels along the base of the brain, and also whether there are, or have been, symptoms which suggest cerebellar disease. Moreover, optic neuritis is very constant and early in its appearance in cerebellar disease, and often late when the tumour is higher up.

I would not have it understood that all tumours in these positions give rise to the phenomena described, for soft succulent ones often do not ; but if the symptoms I have pictured be present, the disease is probably a hard slowly growing tumour in the area mentioned. In all the cases just related, the tumour was a large tubercular mass ; and the early stage of such tumours is shown in a specimen on the table, which was taken from the body of a baby who presented no obvious symptoms of cerebral disease. A caseous nodule can be seen occupying the lower third of the external segment of the lenticular nucleus. In a drawing of a microscopic section of this mass, a large vessel is seen, and close by the vessel lies a colony of well-stained tubercle bacilli.

But, interesting as this series of cases is from the point of view already considered, they are still more so in their bearing upon the question of the functions of the cerebellum, and its relation to muscular spasm. In the 'Lancet' for 1880, vol. i, p. 122, Dr. Hughlings Jackson has written " On Tumours of the Cerebellum," and speaks of " cerebellar rigidity." " I suppose," says he, " the cerebellar rigidity to be owing to unantagonised cerebral influx (rigidity in hemiplegia being owing to unantagonised cerebellar influx), and the tetanus-like seizures to be owing to cerebellar discharges analogous to those cerebral discharges which produce unilaterally beginning (epileptiform) convulsions. But there is the obvious objection that the two symptoms, with the cerebellar disease, may be owing to interference with the corpora quadrigemina (electrical excitation of which produces tetanus-like states), or to interference with the medulla oblongata."

Further on, in the same volume of the ' Lancet,' p. 522, Stephen Mackenzie has a paper on the same subject, in which he expresses the opinion that " the reeling gait and tonic spasms are both expressions of disorder of the locomotory apparatus, of which the cerebellum is the chief governing centre." In comparing the tetanus-like seizures with tetanus, he goes so far as to say, " The range of the two is so much alike as to afford the strongest reasons for the belief that tetanus is an affection of the cerebellum."

Now, if the same rigid contractions and the same tetanic seizures occur in disease of other parts at the base of the brain, they cannot be held to be characteristic of cerebellar disease.

Hughlings Jackson says: " This universal rigidity, as a condition arising from cerebellar disease, I have seen only when the tumour has been in the middle lobe," and again : " In the case of extensive rigidity and tetanus-like seizures, the tumour has been of the middle lobe, and has been large." These remarks really furnish a clue to the matter. I would put the case thus : a tumour of the cerebellum does not produce contractures and tetanic seizures, unless it be in such a position and of such a size as to produce pressure on the pons Varolii or medulla oblongata.

The following case illustrates this point well, and shows that tumours may be present in the lateral lobes, or in the central lobe, and that great distension of the lateral ventricles may exist, without these muscular contractions making their appearance.

Case of Tubercular Masses, both in the Central and Lateral Lobes of the Cerebellum, unaccompanied by Tetanic Spasms.—E. C—, æt. 10, was admitted into St. Thomas's Hospital, under the care of Dr. Bristowe, on September 13th, and died on December 5th, 1877. At the age of seven, the patient had a fit, lasting for about two hours, in which her eyes were fixed and her hands clenched. Ever since that time she had been getting blind. The only other illness she had had was " low

3

fever," two years before admission. About five months
before coming to St. Thomas's she had been suddenly
seized with very severe pain in the head and vomiting,
which lasted, off and on, for a day and a half. These
attacks recurred from time to time for five weeks, during
which her sight became worse and worse, but she never
had convulsions. On examination, it was found that she
could not distinguish light from darkness, and that she
had atrophy of the optic discs. She had no deafness, and
was very intelligent and able to give a great deal of
information about her illness. Her head was somewhat
hydrocephalic in shape, and the forehead prominent.
She kept it fixed, and said it hurt her to let it drop backwards. When seated, the legs appeared to be quite
strong, and it was impossible to detect any weakness in
them when one tried to flex or extend them against her
will. When held by the hands she could walk pretty
well, but when she was left to stand alone she quickly
lost her balance. The same occurred when she tried to
walk, and she seemed to have considerable inco-ordination
of the legs. There was a tendency to fall to the right
side, though she was said previously to have usually
fallen to the left. When standing up and made to lean
a little backwards she did not fall, but ran backwards.

During the rest of her life she had frequent attacks of
severe pain in the head, accompanied by vomiting, and a
feeling of great giddiness. Her temperature remained
normal, and she had no tetanic spasms or contractures.
She was found dead in bed on the morning of December
5th. At the necropsy, there was no evidence of meningitis, but the ventricles of the brain were distended with
clear fluid, twelve ounces being removed and measured.
The convolutions were flattened, and the whole extent of
them collapsed when the ventricles were emptied. The
optic nerves were markedly atrophied. The lateral lobes
of the cerebellum were both very soft and swollen, and
the left was adherent to the dura mater. There were
several tumours, both in the central and in the left lobe,

which appeared to be tubercular masses with the centres broken down. Their size varied from that of a pea to that of a medium-sized marble ; their circumference consisted of a pale, rosy grey, semi-translucent, hard tissue, and their centres of a puriform fluid. The other organs were healthy.

In the class of cases which I have been discussing, tremors and paralysis are probably due to slight interference with the functions of the fibres of the pyramidal tracts, persistent rigidity to still greater interference, while the attacks of tetanic spasms may resemble the convulsions which are observed in connection with cortical tumours of the brain. In the case of cerebral tumours, the pressure is exerted upon the cortical grey matter ; in the cerebellar cases, upon the grey matter in the floor of the fourth ventricle, and not upon that of the cerebellum itself, for I have seen cases in which the central lobe of the cerebellum has been completely destroyed by tumours, but in which no pressure was exerted upon the medulla oblongata, and no rigidity was observed during life.

It is interesting to note, in relation to this question, that Nothnagel has found that irritation of these parts is particularly liable to produce convulsions ; so much so, that he has supposed a special convulsive centre (Krampfcentrum) to exist in the pons, which is readily brought into action, either directly or reflexly, by stimulation of a spot in the fourth ventricle. I have said that the tetanic spasms "may" be due to pressure on the grey matter of the floor of the fourth ventricle, because they are capable of another explanation; they may, in fact, be the result of varying pressure upon the pyramidal tracts. In the next lecture I shall relate a case where a tumour pressing upon the cord from without produced similar spasmodic attacks ; and they are also seen in other diseases affecting the lateral columns of the cord.

Foster, in his work on ' Physiology,' says : " We must consider the cerebellum as an important organ of co-ordination, though we are unable at present to define its functions more exactly." This probably represents our knowledge of the functions of the cerebellum, and it is in keeping with pathological experience ; and I do not think we are justified in considering that disease of it produces spasmodic contraction of muscles, except indirectly, by pressure upon underlying structures.

In my first lecture I considered the part played by chronic diseases, which injure the intracranial portion of the pyramidal tract, in the production of spasm. It was pointed out that the results are similar, whether the lesion is situated in the motor area of the cortex, in the central ganglia, in the crus, pons, or medulla, provided that the function of the fibres of the pyramidal tract be interrupted.

A point which is of great importance, and which will be again referred to when we come to a consideration of functional spasm, is this, that congenital defects of portions of the central nervous system are as capable as destructive diseases of producing paralysis of limbs accompanied by spasm.

Case of Congenital Spastic Hemiplegia.—In the ' Medico-Chirurgical Transactions,' vol. lxvi, 1883, I published " A Case of Asymmetry of the Brain," and discussed the bearing of the anatomical peculiarities of the specimen on the the question of the connection between the optic nerves and certain definite areas of the cerebral cortex. But the case is also a good example of the disease called congenital spastic hemiplegia, and for this reason I refer to it here. E. A— was admitted into St. Thomas's Hospital, under the care of Mr. Mason, owing to injuries received by the fall of a house ; she died of the results of the accident. I never saw her until I performed the post mortem examination upon her ; and no account was obtained during her lifetime of the abnormality in her right arm and leg, nor of the state of her mind and special senses. After death, her father informed me that his daughter had always been healthy up to the time of the accident, which proved fatal

in a few weeks ; that she had never suffered from fits or
other nervous affections; that from her very birth she had
had a small and stiff right arm ; and that she had always
been left-handed and a little lame.   He had never noticed
any mental deficiency ; and she had never complained of
her sight, hearing, or other senses.   No peculiarity in her
eyes, speech, or aspect was noticed during her stay in the
hospital ; but the sister of the ward said that she struck
her as being intellectually dull, even for a person in her
position in life.

At the post-mortem examination, both the right arm and
leg were seen to be smaller than their fellows, though
they were otherwise normally formed ; and the muscles
examined presented the rich brown colour of health.   The
right wrist measured 4¾ inches, the left 5¾ inches ; the
right forearm 6½ inches, the left 8 inches.   The circumference of the left leg, round the calf, exceeded that of the
right by one inch.   The heart, lungs, and other viscera
were healthy, with the exception of the kidneys and the
bladder, which were in a state of acute inflammation.   The
skull and the membranes of the brain were normal, as were
the intracranial blood-vessels, nor was there any marked
inequality of the latter on the two sides.

I need not enter into the minute details which I have
given in the paper referred to, but I may simply state that
both the history of the patient and the anatomical peculiarities of the brain, showed that it was a case of malforma-
ation of one portion of the left hemisphere, and arrest of
development of the remainder.   The most striking charac-
teristics of the specimen which is here shown, were : 1, the
general arrest of development of the left hemisphere, ·
including the motor convolutions ; 2, the small size of the
corresponding crus cerebri and anterior pyramid ; 3, the
absence of the angular gyrus, and superior temporo-sphenoidal convolutions, together with the fusion of some of
the other convolutions of the left temporo-sphenoidal lobe ;
4, the small size of the optic tract, optic thalamus, and
corpora geniculata on the same side.   The spinal cord was

subsequently examined, and, throughout its whole length, it was seen that the right lateral column was smaller than the left, and the left column of Türck smaller than the right.

These differences were most marked in the upper regions of the cord, and became less and less so towards the lumbar enlargement. The microscope shows that in those parts where reduction in size was evident, even to the naked eye, the nerve-fibres were fewer in number and the tissues a little denser than in health, and the general impression left upon the observer's mind was that there was an absence of nerve-fibres, and consequent undue prominence of the connective tissue.

In this case there was clearly congenital absence of part of the motor convolutions, and of the corresponding pyramidal tract; and, clinically, paralysis with spasm was observed. Such a combination seems to prove that the mere absence of the controlling impulses which in healthy subjects traverse the pyramidal tract, is capable of giving rise to a spastic condition of the limbs.

Congenital defects similar to that which has just been described may be bilateral, and produce a condition about which a good deal has been written, and which is termed " infantile spasmodic paralysis," or " spastic paraplegia of infancy." Dr. Hadden, writing on this disease in ' Brain,' vol. vi, 1884, says, " The lower extremities are more or less flexed at the hip- and knee-joints, the thighs are rigid and adducted, the knees in contact, the legs inclined outwards, and the heels often drawn up from the ground by the contraction of the gastrocnemii." Fig. 8 represents a well-marked, but not extreme, case of the kind, which was under Mr. Clutton's care in St. Thomas's Hospital. The child was an idiot, as many of them are. Dr. Ross has proved, by post-mortem evidence, that defects in the development of the brain can produce this condition. In ' Brain,' vol. i, p. 477, he records a case of the disease in which he found at the post-mortem examination a congenital defect in the motor convolutions on both sides, and absence of the large pyramidal cells normally existing in

them.   He found the same condition of the spinal cord as
was present in my case of congenital spastic hemiplegia,
namely, absence of a certain amount of the lateral columns,
but no other abnormality.

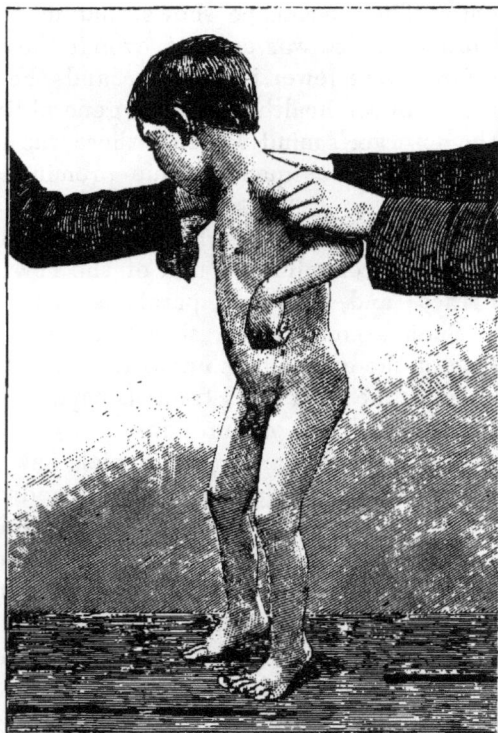

FIG. 8.—From a Photograph of a slight case of Congenital Spastic Paralysis.

Whether all cases of congenital paralysis and spasm of
the limbs be due to arrest of development of portions of
the brain is at present uncertain, but it is not at all improbable that intra-uterine disease of parts already formed
may be the explanation of some.

Contracture, or fixed spasm, is a comparatively frequent
result of cerebral lesions which interrupt permanently the
voluntary impulses proceeding from the cortex.   But, in

addition, mobile spasms have been observed in great variety, either after hemiplegia or independently of it. Sometimes they attend voluntary movements only, sometimes they are involuntary and continuous ; but in all of them, the retention of a large amount of voluntary control over the limbs affected is a striking feature.

In addition to the mobile spasm there is often a certain degree of fixed spasm, though a marked degree of the latter would be incompatible with the development of the former. Continuous movements of the hands and toes have been termed athetosis. The following is a case of the kind which I reported in ' Brain ' for April, 1885. It is a good example of this condition, though rare because of the number of muscles affected, and because the movements appeared after many years on the opposite side of the body.

G—, æt. 24, came under my care as an out-patient at St. Thomas's Hospital on March 26th, 1884. Until the patient was three years old he was quite healthy ; he was then suddenly attacked with loss of power all down the left side, accompanied by repeated and prolonged convulsions. As far as he knew, the movements of the limbs and face of the left side gradually followed the loss of power and had gone on increasing ever since ; at any rate, they had done so as long as he could remember. When between four and five years of age, he was taken to Great Ormond Street Hospital, where his parents were told that the movements were so slight that he would get over them. About three years ago weakness and movements came on in the right leg, and he went to Guy's Hospital, where he was told that nothing could be done for him. He then consulted a homœopathic chemist, who gave him " strengthening medicine," and his right leg became strong again, and the movements in it ceased in three months. A month ago, the weakness and movements recommenced in the right leg, and gradually increased. The arm was said to be free from them. The patient had never had any other severe illness, had not suffered from headaches, and had

done his work well as a carpenter until the last six months.
During this time, the movements of the left hand had been
so bad that he had been unable to hold the nails with
accuracy, though he could hammer well enough with his
right arm. He said he had never had any defect of sight
or hearing.

On rough examination, vision seemed normal, and he
heard the tick of a watch a long way off. There were
movements of the greater part of the muscles on the left
side of the body, and they ceased during sleep ; there was
no anæsthesia. The tongue moved a little irregularly, as
it does in chorea ; speech was drawling. The muscles,
both of the left and of the right side of the face, contracted
too strongly when the patient attempted to speak, so that
a kind of grimace resulted, in which the lines at both
angles of the mouth and the naso-labial furrows were very
deeply marked. There were spasmodic, more or less
regular, movements of the neck to the left, and the shoulder
was rhythmically raised, and jerked slightly forwards and
backwards. One could see the serrations of the left
serratus magnus contracting rhythmically. The abdominal
muscles were harder on the left side than on the right,
but there were no evident movements in them. There
were slight spasmodic movements in the left arm, but they
were only slight. The hand, on the contrary, was markedly
affected. There was constantly more or less rhythmical
flexion and extension of the fingers, mainly at the metacarpo-phalangeal joints, with irregular flexion and extension
of the phalangeal joints, and occasional separation of the
fingers. In the left leg there were similar slight movements of the muscles, with marked movements of the toes,
the great toe being most affected. They were tremulous
flexions and extensions, but the flexor longus pollicis was
constantly contracted, so that the great toe was always
riding over the others, and its proximal phalanx was
usually at right angles to the metatarsal bone. In the
right foot, the movements were similar but not so marked.
In the right hand only the little finger moved, and that so

slightly that the patient had not noticed it. The grasp of the right hand, though stronger than that of the left, appeared to be deficient in power ; the patient said, however, that he had never noticed that it was weak. There was neither atrophy nor hypertrophy of the muscles of the limbs. The patellar reflex was unnaturally brisk on both sides, but there was no ankle-clonus. The patient seemed very intelligent, and said that his memory and mental powers were very good. The fundus of the eyes and fields of vision were normal. The urine contained no albumen.

The next case is one of fixed, combined with mobile spasm, and also differs from that which I have just described in this, that there was no history of hemiplegia or monoplegia, preceding the onset of the muscular contraction.

Case of Mobile and Persistent Muscular Spasm of the Upper Extremity.—E. F—, æt. 11, came to St. Thomas's Hospital as an out-patient, under me, in June, 1884. She was a very well made, intelligent, lively child, and in these respects was said to resemble the rest of her family. When she was nine months old the mother noticed that she held her left arm behind her back, and did not use it properly. Ever since, the arm had been becoming gradually more distorted and useless. On examination, it was found that the left arm was not undersized, or atrophied, but that it was the seat of extreme distortion, which rendered it perfectly useless. The forearm was excessively pronated, so that the palm of the hand looked outwards, and the wrist was flexed ; the metacarpo-phalangeal joints were extended, the proximal phalangeal joints hyperextended, and the peripheral flexed. The elbow-joint was flexed, and looked unshapely. The metacarpo-phalangeal joint of the thumb was flexed, and the phalangeal joint hyperextended ; and the thumb, as a whole, was turned inwards towards the palm. All the parts described were rigidly fixed in these positions, apparently by the spasmodic action of muscles. But the spasm could, with difficulty, be overcome, and then the limb assumed quite a new appearance. Thus, if pronation were rectified and the elbow bent, the wrist at once became extended, all the fingers flexed, and the thumb was turned into the palm of the hand beneath them. This state of affairs was only retained by keeping the elbow-joint flexed, and forcibly overcoming the tendency to excessive pronation. As soon as the latter was allowed to assert itself the whole limb at once returned to the original position. Sensation was unaffected. Dr. Kilner, who tested the electric condition of the affected parts, reported that the muscles reacted normally to the induced current, and presented very slight reaction of degeneration when the constant current was used.

On giving chloroform, the distortion almost entirely disappeared, the muscles were flaccid, and the tissues of the elbow-joint appeared very lax. The child had attended the Orthopædic Hospital for one year without deriving any benefit, but electricity had not been tried. Dr. Kilner was good enough to undertake treatment by this means ; but though he continued it for weeks the condition underwent no improvement.

FIG. 9.—Case of Spastic Paralysis of the left arm and hand, showing the position assumed when the wrist was extended.

Later in the year, on August 24th, 1885, I took her to

have the abnormality photographed, and I then found that certain alterations had occurred in the condition of the hand and arm. The most marked change was that they were never still; in fact the case might then have been

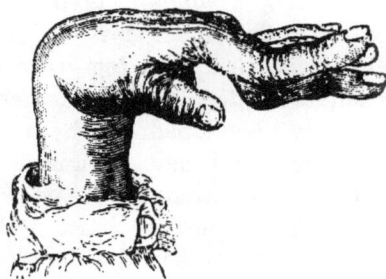

FIG. 10.—Showing position assumed when the wrist was flexed.

called one of athetosis of the left upper extremity. The arm, when left alone, usually occupied a position midway between the perpendicular line of the body and a horizontal line drawn through the left shoulder-joint, and the latter was extended, the forearm pronated, the wrist flexed, and the fingers flexed at the peripheral phalangeal joint, and hyperextended at the proximal. Such a degree of hyperextension now existed in the metacarpo-phalangeal joint of the thumb, that partial luxation had been produced. The child generally liked to hold the hand across the front of the body, with the fingers a little flexed. Irregular clonic jerks affected the whole arm from the shoulder, and the fingers were in more or less constant movement when left to themselves. There were occasional slight jerks of the left angle of the mouth, but the leg appeared free from any abnormality. The child asserted that she had always had these movements, though I certainly did not notice them when I first saw her.

Figs. 9 and 10 represent the condition of things in January, 1886.

Mobile spasms do not always assume the characters which are described under the term "athetosis." Movements resembling, more or less, those of chorea, disseminated sclerosis, ataxy, and paralysis agitans likewise occur.

On June 11th, 1885, a woman, aged 60, came to my out-patient room complaining of loss of power all down her left side; her face, arm, and leg were all weak, and there was a considerable degree of hemianæsthesia. This condition had slowly developed in the course of three months. Though she had had rheumatic fever twice, there was no apparent disease of her heart or other viscera, but her arteries were somewhat hard. By November, moderate hemiplegic rigidity had developed, and, in addition, continuous movements exactly resembling those of paralysis agitans.

It is unnecessary to describe the various forms of mobile spasm which are seen at the bedside. Dr. Gowers has given an account of a considerable number, in a paper which is contained in the 59th volume of the 'Transactions' of the Royal Medical and Chirurgical Society; and Charcot and others have also recorded cases. What concerns us more on the present occasion is the question of their pathology. Can they be attributed to any lesion constant either in its nature or in its position? The answer to this must be in the negative; but we have to confess at the same time that our knowledge of this class of spasm is very deficient. Gowers, in the paper already referred to, records a case of "post-hemiplegic inco-ordination," in which he found, at the post-mortem examination, one lesion only, and that was a cicatricial induration confined to the optic thalamus, and unaccompanied by any pathological changes in the spinal cord. Weir Mitchell found the seat of the lesion in two cases in the corpora striata. In the 'Revue de Médecine' for May, 1883, Émile Demange records the post-mortem examination in nearly a dozen cases of post-hemiplegic mobile spasm, and among these is a case of hemiathetosis, in which there was

extensive softening of the convolutions, but no disease of
the central ganglia. In most of the cases, the lesion or
lesions were in the central ganglia, sometimes involving
the internal capsule and sometimes not. Hence it seems
to me that we must agree with Demange when he says
that these movements may occur from disease of any part
of the motor tract in the brain ; and he adds that the
variety of movement does not depend upon the seat of the
lesion, but rather upon the condition of the motor tract
below the lesion.

All the varieties of mobile spasm seem to result from a
mixture, in varying proportions, of paralysis, spasm, and
irritation ; and their development depends upon lesions
which interfere with the perfect functions of the motor
centres and fibres, but which do not completely interrupt
them. That such a condition of incomplete destruction is
competent to give rise to movements of the kind under
consideration is well shown by a case which I published
in the 'Lancet,' in a paper on Cerebral Localisation
already referred to. The patient was a woman, aged 46,
who was admitted into St. Thomas's Hospital on July
20th, 1878, and died on July 21st, 1879. She was suffering from chronic renal disease, and on January 18th,
1879, she complained of being unable to keep her left arm
still. The movements both of the arm and leg soon
became violent and disorderly. When asked to put her
left hand to her face she did so hesitatingly, and with a
series of jerks which had no constant rhythm or direction.
The same occurred on voluntary movement of the leg,
while the right arm and leg moved slowly and naturally.
The jerking movements gradually subsided, until, on
January 30th, the leg was freely and normally moved,
while the left arm was almost completely paralysed ; and
although it recovered some degree of power, it remained
markedly paralysed to the end of her life. After death,
the middle part of the ascending convolution on the right
was softened, shrunken, and of an orange colour. It is
clear, in this instance, that the involuntary movements of

the limb occurred at the time when the cortical grey
matter was in the early stages of softening, and that they
ceased when paralysis resulted from its complete disorga-
nisation.

In concluding the consideration of spasm in connection
with diseases of the intracranial portions of the voluntary
motor path, it is necessary to refer to a statement which I
made in my first lecture; namely, that "our present
knowledge of anatomy, physiology, and pathology does
not justify us in concluding that there is any efferent
motor connection between the brain and the spinal cord,
except the pyramidal tract, direct and crossed, disease of
which gives rise to chronic muscular spasm. Evidence in
support of this view will appear in subsequent portions of
these lectures."

In saying this, I wished to exclude the cerebellum and
the central ganglia from consideration, since any rigidity
or spasmodic contraction of muscles which arises in connection with diseases of the areas they occupy seems to
depend upon involvement of the pyramidal tract, which
passes close by them. Evidence has already been supplied in support of this view, as far as the cerebellum is
concerned. Let me now record a case which shows that
destruction of the central ganglia produces no spasm, so
long as the neighbouring internal capsule is intact.

C. B—, æt. 39, had an attack of right hemiplegia ten
years before death, and another somewhat later. He died
of carcinoma of the cæcum, having long recovered completely from all paralytic symptoms. At the post-mortem
examination I found that the hemispheres were free from
disease. On exposing the ventricles it was at once seen
that the left caudate nucleus was very much shrunken,
and of a yellowish hue. Both optic thalami and the
right nucleus caudatus looked healthy. The left crus
cerebri was small and discoloured along its inner third, and
the left side of the pons Varolii was evidently flattened.
Nothing abnormal was noticed in the crus cerebelli on
either side, and the pyramids in the medulla oblongata

were equal and natural.   No descending degeneration
was to be discovered in the spinal cord with the naked
eye.

The brain was then set aside to harden in alcohol,
before any further examination was made.   A series of
horizontal sections through the hemispheres and central ganglia showed that there was hardly anything left
of the nucleus caudatus, or of the nucleus lenticularis on
the left side, but there were no evident changes in either
division of the internal capsule.   The optic thalamus was
a little smaller than its fellow on the opposite side, while
all the central ganglia on the right were healthy.   A vertical section through the crura cerebri showed very marked
atrophy of the left as compared with the right.   A similar
diminution in size was observable in the left half of the
pons ; but the pyramids, and all other divisions of the
medulla oblongata, were healthy and symmetrical.

This case shows that disease so severe as to result in
almost total obliteration of the corpus striatum, may produce hemiplegia, which passes off without leaving rigidity,
paralysis, or any other symptom behind.   And, secondly,
it proves, as far as any single case can prove, that the
efferent fibres of the corpus striatum do not pass down
further than the pons Varolii or medulla oblongata.

The continuous degeneration of the lateral column
through the whole length of the spinal cord in cerebral
lesions shows that the fibres are not interrupted by enter-
ing ganglionic centres, but pass straight to their several
destinations in the anterior cornua.   It will be readily
understood that, if disease destroy them at any level in
the spinal cord, degeneration will occur in them below, and
give rise to symptoms of hyperphysiological activity of the
spinal centres with which they are connected, similar to
those which occur when the disease is intracranial.   In
the latter case, all the spinal centres which are under
voluntary control are affected ; in the former, only a certain number, more or less numerous according as the
lesion is situated high up or low down in the cord.   There

4

is this difference, too, that, in cerebral disease, the rule is
to find the contracture unilateral, in lesions of the cord bilateral. But this is merely because the pyramidal tracts
in that part of the brain where they are most liable to be
interrupted by disease, lie at some distance from each
other ; whereas, in the spinal cord they are so close
together as usually to be involved in the same pathological changes. The symptoms, however, of disease of
the lateral columns of the cord are the same, whether it
follows cerebral or spinal lesions. The affections of these
columns are either primary or secondary, and the symptoms differ slightly accordingly. Thus, if a primary
chronic inflammation attack them, as in primary lateral
sclerosis, or disseminated sclerosis, paresis of the corresponding muscles is an early symptom ; and tremors of
the limbs when moved, increased tendon-reflexes, and
clonus soon follow. In addition to these, transitory involuntary contractions, and longer, though still inter-
mitting, spasms occur, and finally culminate in persistent
rigidity. When the latter has fairly set in, the muscles
can only be loosed from their slavery by a process which
condemns them to perpetual inaction. Their masters, the
cells in the anterior cornua, may be attacked and die ;
but the muscles perish with them, and the tendon-reflexes
and other signs of their activity disappear for ever. But
the lateral columns are more often affected secondarily
than primarily, and then the symptoms which culminate
in spasm follow those which characterise the original
disease. Thus we often see cases of myelitis run their
course to destruction of a portion of the spinal cord,
and rigid contracture of the limbs then supervenes.
Such disease occurs more frequently in parts below the
cervical regions than in those above it, and consequently
the legs are more often the seat of rigidity than the
arms.

But matters are not always so simple as this, for a
lesion of the lateral columns may be combined with that
of some other tract of fibres ; and a combination of sym-

ptoms then presents itself, some being attributable to disease of one tract, and some to disease of the other. If we are to draw correct conclusions with regard to the localisation of diseases of the spinal cord and their progress, we must never lose sight of the phenomena which characterise the healthy action of the nervous system. Physiology is the only safe guide for the diagnosis of the positions occupied by disease. Again, although pressure upon the pyramidal fibres interrupts their functional activity, the symptoms are very different according to the manner in which the pressure is applied, whether from without or from within the cord. As time will not permit me to enter into detail upon all those diseases of the spinal cord which give rise to spasm, I shall try and illustrate from my own experience the two points to which I have specially referred; namely, first, the difference in the results of pressure applied upon the cord from without and from within ; and secondly, the modification which occurs in the symptoms of disease of the lateral columns, when it is combined with disease of other parts of the cord.

The following case is a good instance of pressure on the spinal cord from without, giving rise to interference with the function of the lateral columns, and to extreme contracture of the lower limbs.

Case of Tumour (Myxoma) on a Spinal Nerve, which pressed upon the Cord, and produced Paralysis and Spasmodic Contraction of the Lower Limbs.—A. J—, æt. 24, blacksmith, was admitted into St. Thomas's Hospital, under the care of Dr. Bristowe, on April 20th, 1876. His family history was unimportant, and, until his present illness began, he had always enjoyed good health. He had drunk freely both beer and spirits. He was in the habit of lifting heavy weights, and had occasionally "ricked" his back, but had never suffered from any serious injury. In August, 1875, he began to feel a sensation of weakness in the loins, but he had no actual pain. This was followed in November, 1875, by a certain degree of weakness in walking, which slowly increased, so that he became unable to walk at all two months before admission. During this period he suffered from cramp, with a slight loss of power in his left arm, and from occasional attacks of sharp, shooting pains and cramp in his legs. He had had no loss of sensation or numbness anywhere.

On admission, he was found to have marked paraplegia, but no loss of sensation. He suffered from cramps, and his legs at times were flexed involuntarily. He had incomplete control over his evacuations. His left hand and arm were not as powerful as his right. All his organs appeared to be healthy, and he had no albumen or pus in his urine. Somewhat later, painful contraction of the adductor muscles of the thighs used to come on and last for hours, and he had slight numbness as high as the umbilicus. When these attacks of spasm supervened in the muscles of the leg, pain and similar cramps were experienced in the left arm. The remainder of his life was very painful. His legs became gradually more and more permanently and rigidly flexed and adducted, and bed-sores developed over the internal aspects of the knees, where they were forced against each other, as well as over the sacrum and trochanters. Sensation, however, rather improved than deteriorated. Finally, incontinence of the evacuations, cystitis, and hectic fever set in; and the patient died on August 17th, four months after admission, and one year from the date of his first symptoms.

Post-mortem examination.—Most of the organs were found to be perfectly healthy, the urinary tract and spinal cord alone presenting evidences of disease. The pelves of the kidneys and the ureters were inflamed, and the kidneys were large and succulent. On one of the anterior spinal nerves, which originated just at the termination of the cervical enlargement, was a tumour about as large as a medium-sized marble. It was firm, pale in colour, and covered with a perfectly smooth fibrous sheath. The nerve was lost in its substance, but the tumour was free from adhesions with the spinal cord and all the surrounding structures. It was wedged in between the bones forming the spinal canal on the one hand and the anterior surface of the spinal cord on the other, so that the latter was flattened by it anteriorly and slightly laterally. The tumour was a myxoma.

The motor phenomena which characterised this patient's history were paralysis, attacks of tetanic spasm, and finally, persistent rigidity of the legs. Making allowance for the position at which the pressure was applied to the spinal cord, namely, below the origin of the nerves supplying the arms, the occurrence of tetanic spasms in the legs from time to time reminds us forcibly of similar but more extensive spasms which occur when the disease occupies the base of the brain, or the cerebellum, and, I think, gives support to the conclusions which were drawn as to the part played by the cerebellum in their production. A point which should be specially noted in this case is this, that when the pains and tetanic spasms occurred, pain and spasm were likewise felt in parts supplied by the nerve on which the tumour was situated. These attacks of spasm in the legs, which were intermittent, appear then to have been produced by changes occurring in the tumour, which also showed their presence by increased pressure on the nerves for the arm which it involved. The same intermittence was observed in the spasmodic attacks produced by tumours which occupied the base of the brain, and the same thing occurs when the tumour presses upon the cerebral cortex. We may conclude, therefore, that in all such cases the convulsions are due primarily to alteration in the condition of the tumour, though we cannot say in what the alteration consists.

The case I shall relate next is a good instance of the results of pressure on the cord applied from within by central tumours. The disease is a very rare one, and the specimen I show is so beautiful an example of it, that I have endeavoured to work out the histological changes which it has produced in the cord as minutely as possible, in order to see how far they explain the clinical phenomena observed during life.

C. W—, æt. 50, waiter, was admitted into St. Thomas's

Hospital, under the care of Dr. Stone, on January 24th,
1885, and died on March 18th, 1885.   His family history
was unimportant; he had a wife and five children, all
strong and well.   He had enjoyed the best of health since
he was a child, and had never had syphilis.   His illness
commenced somewhat suddenly, with pain in the back,
eighteen months before admission.   Partial loss of power
and sensation in both his arms and hands soon followed,
together with tremulousness on attempting to use them.
It was not until three months before coming to the hospital
that his legs became affected.   Since then he had had
numbness and pricking sensations, loss of power, and a
sensation of coldness in them.   On walking, the ground
felt soft.   His bowels had been irregular, and he had had
some difficulty in passing urine; his appetite had been
poor; he slept fairly well.

On admission, he was a healthy-looking man, and complained of loss of power, numbness, and formication in his
legs, together with incontinence of urine.   On examination,
no angular curvature or marked rigidity of spine was
discovered; but some tenderness, without prominence, was
found about the lower cervical and upper dorsal vertebræ.
The legs were almost completely paralysed and anæsthetic,
but there was no wasting of muscles, no rigidity, and no
tremors.   The tendon-reflexes were much exaggerated,
and there was ankle-clonus in both legs.   The patient
could use his hands and arms fairly well, but they were
numb and tremulous, and the grasp was weak.   There
was no ocular paralysis.   Urine was retained, and the
bowels confined.   Pulse was 84; temperature normal;
tongue slightly furred.   The urine was of specific gravity
1030; no albumen; faintly acid.   The abdomen was
natural; hepatic and splenic dulness was normal.   The
lungs were resonant, and the breathing healthy, nor was
there anything abnormal detected on examination of the
heart.   He continued in much the same condition for
some weeks, complaining a good deal of pain in the back
high up, and of occasional pains in the legs.   The muscular

power and sensation underwent but slight variations in the
upper and lower extremities, though he complained sometimes of twitchings. The temperature was generally
normal, but now and then somewhat raised. Incontinence
of urine set in about February 5th, and it became alkaline
and phosphatic, and contained albumen.

On February 24th, it was noted that there was tenderness in the lower dorsal and lumbar regions of the spine,
and that complete anæsthesia existed below the level of
the middle of the chest.

On March 12th, the legs were completely powerless, and
the right leg œdematous. The patellar and plantar reflexes
were very brisk ; ankle-clonus was marked, but there was
no knee-clonus. Sensation of touch and of pain, in lower
extremities, was abolished ; sensation of heat was retained.
There was no control over the bladder or rectum. In the
centre of the sacral region was a large horseshoe-shaped
bedsore, with a red granulating surface, and a smaller
one was forming over the right hip. The urine was alkaline, and contained pus and mucus.

On March 14th the patient took his tea at 5 p.m., and
seemed to be in his usual condition. Soon after 6 p.m.,
when the sister came to him, he was found to be unconscious ; his eyes were shut, and he was breathing heavily,
and when asked how he was, he said he was " as usual."
The pupils acted to light. His tongue was protruded to
the left, his left cheek was flaccid, and his left arm powerless. His head and eyes were turned to the right. The
pulse and heart-sounds were very feeble, but there was no
murmur. He altered but little after this, and died on
March 18th.

The following is an account of the post-mortem examination, which I made twelve hours after death :—Body
emaciated. Over the sacral region in the middle line, a
large superficial bedsore was seen and there were several
smaller ones in the neighbourhood. The peritoneum, peri-
cardium, and right pleura were healthy ; the left pleura
was everywhere firmly adherent to the thoracic wall. The

lungs were intensely congested and œdematous, otherwise
healthy. The heart appeared normal in size, the valves
on the right side were natural, those on the left a little
thickened, but competent. Along the line of attachment
of the mitral valve to the wall of the heart, on its auricular
aspect, was a series of papilliform elevations, such as are
often found in this position in cases of endocarditis. There
appeared to be little or no coagulum upon their surface.
The spleen was large, weighing $9\frac{1}{2}$ ounces, and contained
several large pyramidal infarcts, which were still soft, and
had a very clearly defined white margin.

The bladder, ureters, and the pelves of the kidneys
were inflamed and presented some phosphatic deposits
upon their surface. The kidneys were somewhat large,
succulent, and showed evidence of early suppurative ne-
phritis. The capsule peeled off easily, and below it were
seen, in large numbers, minute white dots, which felt
gritty when touched with the point of the knife. In one
kidney was a small infarct, opaque yellowish white in
colour, with a red margin. The weight of the kidneys
was $10\frac{1}{4}$ ounces.

The liver was somewhat large and very heavy, weighing
$70\frac{1}{2}$ ounces. The pyloric end of the stomach was attached
by slight adhesions to the gall-bladder. The latter was
completely filled by a large gall-stone. Its walls appeared
to be fairly healthy except at the extreme fundus. The
latter was thickened by new growth and projected as a
hard nodule in the small indentation, through which the
healthy gall-bladder often projects on the upper surface of
the liver. From the nodulated fundus as a centre, grew
a large mass of carcinoma into the substance of the right
lobe of the liver. The line which limited its growth
farthest away from the gall-bladder was curved, and in its
neighbourhood the luxuriant new growth bulged upon the
surface of the organ. That portion which was nearest to
the gall-bladder was depressed and harder, giving one the
impression that it was older than the peripheral growth.
Large vessels coursed over the surface of the mass. One

or two very minute islands of new growth were seen here
and there in the liver-substance at a distance.

The stomach and intestines were healthy and no focus
of new growth was found which could have given rise to
that in the liver as a secondary product. The spinal
column, as seen from the interior of the body, seemed to
be perfectly natural; nor, on removing the posterior portion of it and taking away the spinal cord, did it present
any traces of disease.

On removing the brain, the greater bulk of the right
hemisphere, as compared with the left, was very striking ;
to the finger it felt soft, especially in the lower part of
the parietal region and in the temporo-sphenoidal lobe.
On making a horizontal section so as to expose all the
central ganglia to view, it was seen that the whole of the
corpus striatum—both the nucleus caudatus and nucleus
lenticularis—was softened and broke up under a stream of
water. The internal and external capsules, the convolutions of the island of Reil, and other convolutions adjacent
to them were in a similar condition. The right middle
cerebral artery was found to be completely blocked, just
beyond the spot where the first few vessels are given off
to the central ganglia.

On examination of the spinal cord, it was found that
the dura mater was healthy. At the junction of the lower
dorsal and upper lumbar regions was a softened area about
one inch in length and involving apparently the whole
thickness of the cord. It was rather deep red in colour,
with a pale yellowish island at one spot, which, with its
deeper coloured periphery, was very suggestive of infarction
as it is seen in other organs. Higher up, above the middle
of the dorsal region, the anterior spinal artery was interrupted in its course and surrounded by a long superficial
hæmorrhage of a brownish red colour. The cervical enlargement, especially in its upper part, seemed to be of
enormous dimensions, retaining nevertheless the natural
shape of the cord. On pressing with the finger it was
found to be very firm over an area one to two inches in

length.   On making a transverse section through the cord
in the upper cervical region, where there was no feeling
of increased density, the central canal was found to be
much enlarged and to contain blood partially coagulated.
Where the canal opened out into the fourth ventricle, it
appeared to be healthy.   When a section was made through
the enlarged and hardened portion of the cord a most
curious appearance presented itself.   A very narrow and
even rim of apparently healthy white matter surrounded,
as with a sheath, a dense reddish grey mass, of oval outline.   In the centre of this mass was a hollow space, which
presented a very marked resemblance in shape to the central
grey matter of the spinal cord and which contained blood.
On teasing out a portion of this new growth in the fresh
state, it was found to consist mainly of spindle-cells and
of cells with many long fine processes, such as one sees in
gliomata.

In the first of these two cases of spinal tumour, where
the pressure was exerted upon the cord from without,
death occurred in twelve months from the appearance of
the first symptoms.   The latter were, in order of time,
weakness of the legs in walking ; increasing weakness of
legs, accompanied by cramp and shooting pains ; paraplegia, cramp, and involuntary flexing of legs ; slight
numbness and long-continued attacks of muscular spasm
in the lower extremities ; and, finally, persistent rigidity.
In the second case, the symptoms due to pressure alone,
which was exerted upon the cord from within in the
cervical region, continued for fifteen months uncomplicated
by other disease ; they were, simply, partial loss of power
and sensation in the arms and hands, and some tremor in
them when used.   This was the condition of the arms
when the patient was admitted into the hospital ; and it
was little altered even at the time of his death, which
occurred twenty months after the appearance of the first
symptoms.   During the last five months of life, fresh
symptoms arose in relation with hæmorrhage and softening
in the dorsal region of the cord—accidents which are very

liable to occur in cases of gliomata ; and the accident which finally proved fatal was embolism of the right middle cerebral artery, giving rise to left hemiplegia.

When a tumour grows in the spinal canal outside the cord it may produce but few symptoms until it presses the cord against the resisting walls of the canal ; but, after this has taken place, the course of the disease is naturally very rapid, as the cord is quickly flattened by the constantly increasing demands for growing space which are made by the tumour. When a tumour arises, on the other hand, within the spinal cord, it disturbs its functions even from the very commencement ; but, as the nerve-substance appears to be elastic, and to allow a good deal of gradual stretching without serious interference with its functions, a tumour may go on growing for a long time before it produces striking pathological phenomena, either by pushing the cord against the bony walls of the spinal canal, or by exhausting the elasticity of the membranes which envelope it. These remarks only apply to gliomata and such non-malignant tumours as exert pressure on surrounding tissues, but do not invade them. The description which I now give of the microscopic appearances of the spinal cord and of the tumour in the present case, shows that a state of things is produced by the disease which fully explains the chronicity of some of these cases, and the trivial symptoms which present themselves.

The tumour is situated almost entirely in the cervical region, and mainly in the cervical enlargement. In the middle of the latter, the mass assumes its largest propor-

FIG. 11.—Transverse section (natural size) of Spinal Cord in a case of Central Glioma. *a.* Rim representing Spinal Cord. *b.* Anterior Cornu. *c.* Posterior Cornu. *d.* Central Tumour.

tions, and the enveloping sheath of nerve-tissue is comparatively smaller than in any other region of the cord
(Fig. 11).

When examined with the microscope, the greater part
both of the white and of the grey matter can be readily
recognised, though pressure has produced some alterations
in them. On holding up a stained transverse section to
the light, one observes in its anterior third a small oval
area which is more transparent than the parts around; it
is situated in the white rim, and lies upon the tumour.
Under a low power, this proves to be the anterior cornu,
containing the usual multipolar cells. Symmetrically
placed upon the opposite side of the tumour is a similar
oval mass, which is the other anterior cornu. The cells
are not healthy on either side, but are pigmented, and
more or less elongated by pressure, but, considering the
pathological conditions to which they have been subjected,
they are singularly well preserved. The anterior roots
can be traced from the cornua through the white
substance to the periphery of the cord, and the posterior
cornua extend backwards along the margin of the tumour.
The lateral, posterior, and anterior columns of white fibres
are all present, and show but slight pathological alterations; there is no marked sclerosis anywhere. The same
may be said of the whole extent of the cord which is
occupied by the main bulk of the tumour. The latter
becomes smaller and smaller in the lower cervical region,
and is no longer evident as a mass in the lower dorsal
region. Here, however, considerable hæmorrhage and
softening have occurred; and to these changes the later
symptoms in the case may be referred, when the legs
became paralysed, and increase of the tendon-reflexes and
ankle-clonus appeared.

As the tumour becomes smaller, and finally ceases to be
recognised as such in the dorsal region, it is probable that
the changes present in the cord there represent the early
stages of its growth. They occupy the left half of the
posterior columns and the left lateral column, and consist

of a peculiar thickening of the neuroglia, which is trans-
formed into a hyaline substance. The nerve-fibres
in its meshes are at first but little altered, then they
become swollen ; the vessels are much thickened, and their
walls assume the same glassy appearance as the neuroglia.
In the cervical region, the central canal cannot be recognised ; but in the lower parts of the cord, where the
growth is early, it can be seen pushed to the right.

As to the tumour itself, it consists of numbers of islands
of new growth situated around large blood-vessels as
centres, and starting from them. The vessels have the
thickened hyaline walls already referred to, and numbers
of fibrillæ can be seen issuing from them. The new
growth is richly nucleated, and has a very definite radiate
arrangement round each centre. On examining very fine
sections under high powers, the new tissue is seen to
consist of meshworks of very delicate fibrils, and bundles
of fibrils, in which are situated the nuclei. It may therefore be put down as a glioma.

In Virchow's ' Archives ' for 1884 a very similar case
is recorded by Dr. Reisinger, and an abstract of all the
cases to be found in medical literature is added. They
amount to twenty in all. It is very curious that, in Dr.
Reisinger's case, the earliest changes were seen, as in mine,
in the posterior columns near the central canal.

The minute anatomy of the specimen I show here
explains why the symptoms in cases of central tumour of
the cord may be so slight and extend over such long
periods of time, and why spastic conditions are not readily
induced, as they are when pressure is applied from without. For, even when the disease is so extreme that the
substance of the cord forms merely a narrow ring round
it, the nerve-tissue presents very slight pathological
changes.

I shall now relate a case which illustrates the modification which occurs in the symptoms of disease of the
lateral columns, when it is combined with disease of other
parts of the cord.

Case of Primary Disease of the Lateral Columns, accompanied by similar changes in other parts of the Central Nervous System.—M. H—, æt. 53, whose occupation was pulling down old houses, was admitted into St. Thomas's Hospital, under the care of Dr. Ord, on November 10th, 1884, and died on December 2nd, 1884. There was no history of nerve disease in his family. He had always enjoyed fair health, though troubled with indigestion and bilious attacks. The only definite illness he had had was ague at the age of twenty-three. He had never had syphilis. Before his present illness began he had been in the habit of drinking five or six pints of beer daily, but no spirits; he also smoked a good deal, about half an ounce of tobacco a day. His occupation necessitated pretty constant exposure to cold and wet, and he had had several falls, but had never had any symptoms of importance after them. About seven years before admission to St. Thomas's he had had great anxiety about money matters, and was up night and day for about a month. He then got out of sorts, and presently noticed that the toes of both feet were becoming numb and the legs somewhat stiff; this, together with weakness, produced difficulty in walking, and the right leg began to drag. He was quite unable to walk in the dark, or with his eyes shut, or to turn round quickly. The difficulty in walking had steadily increased, and both legs had become equally affected. Six months ago, total inability to get about made him take to his bed, the numbness increased, and he began to be troubled with jerking of the legs; the arms got also somewhat stiff. He noticed that, although he could grasp large objects well, he could not pick up small ones, such as pins. He had burning pains in the legs and shoulders, and shooting pains in the back and abdomen. During his whole illness, he had suffered from indigestion and constipation; his legs had become thinner, and he had lost weight. His sight, which used to be very good, had gradually become dim, so that he could only read print when it was held very close to him. His hearing had remained good. The right foot and ankle had been swollen for a short time a year previously, but none of his other joints had been affected. He had lost all sexual desire, and for five years had had some trouble with his urine, sometimes having retention, at others continuous dribbling. He had never had any difficulty in speaking or in swallowing.

On admission, the patient was an emaciated man complaining of inability to walk, dimness of sight, and shooting pains in the legs and thighs. Although unable to stand, he moved his legs freely in bed and there appeared to be no loss of power in them, but there was some rigidity. The arms showed no loss of power; the grasp was strong with both hands, but there were slight tremors in the right arm when he used it. The muscles of the face were stiff and tremulous when strongly contracted. There was partial loss of control over the bladder, but none over the rectum. Sensation was impaired in the lower extremities, but not markedly; the patellar reflex was absent and there was no ankle- or knee-clonus. Plantar reflex was brisk on both sides, the cremasteric absent. The lower abdominal reflex was slight on both sides, the epigastric absent. The right pupil was larger than the left, and both acted to light and accommodation. There was no oculo-motor paralysis or colour-blindness; he had slight lateral nystagmus. There was well-marked atrophy of both discs, most marked on the inner side.

There was no affection of speech or of hearing, but the sense of smell was much impaired. The radial vessels were somewhat thickened, the bowels constipated. The abdominal and thoracic viscera were free from serious disease. The urine had specific gravity 1025, was alkaline, and contained phosphates but no albumen.

During the rest of his life little alteration occurred in the symptoms which have been described, but cystitis with high temperature set in. He was also attacked with swelling, redness, and œdema of the metatarso-phalangeal joints on both sides and of the ankle-joints, and finally got bedsores, and died on December 2nd.

The following is the account of the post-mortem examination, which I made on December 3rd.   The body was
that of a well-nourished, grey-haired old man.   The legs
were thin and proportionately thinner than the arms.   Both
heels were black as if from commencing gangrene.   The
external malleolus on the left side, as well as the sacrum,
was in a similar condition.   The ankle- and knee-joints
were quite healthy.   The heart, lungs, liver, and spleen
presented no evidences of disease.   The kidneys were
healthy, but the bladder was evidently inflamed, as it contained three or four ounces of flaky, purulent fluid, and its
mucous membrane was of a deep purple colour.   The
intestines, pancreas, and suprarenals were all free from
disease.   The sheath of the spinal cord was distended with
clear fluid, so that when the latter escaped the cord was
seen to occupy but a small part of the tube formed by the
dura mater.   There was no undue vascularity nor anything
remarkable on external examination.   On making transverse sections of the cord from below upwards it was found
that in the lower part of the lumbar region there were no
abnormal appearances.   But in the upper lumbar region
three tracts of degeneration were visible.   Two of them
were in the lateral columns, which were symmetrically
affected ; the third consisted of the columns of Goll, which
together formed a triangular area.   The diseased parts
were transparent and had the colour of glue.   The degeneration could be followed in all three tracts as far as the
medulla oblongata, but there it became less and less evident
and could not be seen at all in the upper part of the medulla
and pons Varolii.   The disease in the lateral columns
appeared to be superficially situated and was more extensive in the upper than in the lower regions of the cord.
It occupied a larger area on the surface of the lateral
columns, but penetrated less deeply into them than is
usually the case in descending degenerations.   It was impossible to say, from a naked-eye inspection, whether the
columns of Goll were the only parts of the posterior columns
which were diseased or whether the degeneration affected

those of Burdach as well. The cord was firm, and, with the exception of the regions already referred to, it appeared to be healthy.

The only morbid appearances discovered in the brain were found in the optic nerves. They were both pale and flattened and evidently the seat of degeneration. The corpora geniculata, corpora quadrigemina, optic thalami, and supposed cortical centres of vision were all normal in appearance. There were symmetrical ivory exostoses on the fourth and fifth ribs on each side in the posterior third of their length, but they were not so situated that they could press upon either nerves or vessels.

The symptoms in this case which point to lateral sclerosis are weakness of the limbs, tremulousness, jerkings, and rigidity; while numbness, pains, loss of sexual desire, difficulty with the evacuation, dimness of sight, and absence of the patellar reflex, indicate more extensive disease.

On microscopical examination, the spinal cord proves to be the seat of very widespread changes; and yet the parts affected are so constant throughout that the affection cannot be looked upon as an indiscriminate one, but must be placed among the more complicated combined system-diseases. The regions affected are (1) the lateral columns; (2) the posterior columns; (3) the direct pyramidal tracts; (4) the anterior root-zones; and (5) in a minor degree, the grey matter of the anterior cornu.

But it will also be observed that, in the neighbourhood of the grey matter everywhere, a zone remains healthy, and, in addition, a superficial tract of fibres situated between the crossed pyramidal and cerebellar tracts posteriorly, and the anterior root-zones anteriorly, is unaltered. The pathological changes throughout the diseased areæ are the same, and must be considered to be those of chronic inflammation affecting mainly the nerve-fibres, and in a far less degree the neuroglia; in other words, they represent a form of chronic myelitis. In the regions referred to are seen increase in the nuclei of the neuroglia, and in the white corpuscles within the vessels, hyaline

5

swelling of the neuroglia, distension of the neurilemma, and disintegration of the white substance of Schwann.

Motor phenomena were those which were most strikingly affected during life, and sensory changes were far less marked. The pyramidal tracts, direct and crossed, and the anterior root-zones, are undoubtedly motor regions, while the columns of Goll are afferent, and connect centres below with centres higher up ; hence the disease is principally an affection of the motor systems of fibres. The escape of the small superficial area in the lateral column from pathological changes is exceedingly interesting, and marks it out as a tract containing some particular set of nerves. Gowers first called attention to this small tract of fibres which appear to be afferent, and to degenerate centripetally. Several other writers have also mentioned it, among them Dr. Hadden, and Dr. Howard Tooth has a paper on it in the ' St. Bartholomew's Hospital Reports ' for 1885, vol. xxi.

The medulla oblongata, pons Varolii, cura cerebri, motor convolutions, anterior crural and optic nerves, and the triceps femoris muscle, were all examined microscopically. The optic nerves were found to be atrophied, and there was an excess of nuclei in the anterior crural nerve and in the triceps femoris ; but with these exceptions, all the parts examined were free from disease.

I have detailed this case, rather than one of the more usual ones, where chronic affections are confined to less extensive systems of nerve-fibres, in order to show that disease is a very complex process, however much we may endeavour to simplify our notions of it, and likewise to point out in how modified a form disease of the pyramidal tracts presents itself, if there be in addition pathological changes in other regions of the cord.

Before leaving the subject which I have thus far been illustrating, namely, the relation between diseases of the cerebral motor system and the production of chronic muscular spasm, a few words must be added in answer to the question, Does spasm ever result from afferent

stimuli reflected on to efferent fibres in the cerebral
centres, or from diseases of the commissures ? If we limit
the term cerebral centres to the grey matter of the hemispheres and central ganglia, I know of no undoubted
instances of reflex spasm in connection with them. Convulsions, which may be due to such causes, do not come
within the scope of the present lectures; and had I the
time to treat of them, I should shrink from casting a
shadow upon ground which has been so brilliantly illumi-
nated by the genius of Hughlings Jackson. Several cases
of tumours of the corpus callosum have come under my
notice, and they have been published by Dr. Bristowe in
' Brain,' vol. vii. I cannot do better than quote the words
of so distinguished a physician and pathologist in describing
the symptoms which were observed in them. The chief
characteristic features were :

" 1. Their ingravescent character, a character which
they possessed in common with other cases of cerebral
tumour.

" 2. The gradual coming on of hemiplegia for the most
part resembling in its distribution the paralytic symptoms
usually attending hæmorrhage into one of the hemispheres,
or softening due to embolism.

" 3. The association with the paralysis of one side of
vague hemiplegic symptoms of the other.

" 4. The supervention of stupidity, associated for the
most part with extreme drowsiness, a puzzled inquiring
look when awake, a difficulty of getting food down the
throat, and cessation of speech.

" 5. The absence of implication of the oculo-motor
nerves, and of direct implication of other cerebral nerves ;
and lastly, death by coma."

From beginnning to end of these cases no muscular
spasm occurred.

In the few remarks which the time at my disposal has
allowed me to make on diseases of the pyramidal tract,
and the part played by it in the production of spasm,
I have tried to point out how numerous and varied the

clinical phenomena may be, according to the level at which
the pathological conditions occur, according to the method
of attack—whether from within or from without—and
according to the presence or absence of disease of other
portions of the central nervous system.

I must now leave this division of my subject, and pass
on in the last lecture to a consideration of the spinal
motor system, and to so-called functional spasm.

LECTURE III.

In my second lecture, I finished the remarks which I had to make on the connection between the cerebral system of fibres and the clinical phenomena of spasm, and I shall now proceed to a short consideration of the spinal system, and the relation between diseases which affect it and spasmodic contraction of muscles.

By the spinal system I mean all those nerve-centres which occur between the central ganglia of the brain and the termination of the spinal cord. Each of these centres consists of afferent and efferent nerves and ganglionic cells; and as my object is to treat the subject from a general point of view, we may divide it into: 1. Spasm produced by diseases of efferent spinal nerves; 2. Spasm produced in a reflex manner by diseases of afferent nerves; and, 3. Spasm produced by diseases of the ganglionic cells.

1. *Spasm produced by Diseases of Efferent Spinal Nerves.*—A very large proportion of the cases in which spasm of muscles produces distortion are not to be explained by supposing that irritation of the diseased nerve gives rise to excessive muscular contraction. On the contrary, the healthy muscles are those actively concerned, while the diseased muscles are for the most part passive. Progressive muscular atrophy is a well-known instance of this, where a peculiar claw-like distortion is produced by atrophy of the interossei and the continued activity of other muscles (Fig. 12). Here the neuro-muscular disease appears to owe its origin to an affection of

the large cells in the anterior cornua of the spinal cord. But even where no gross disease of nerve or muscle exists, it seems to be a rule that, when the origin and

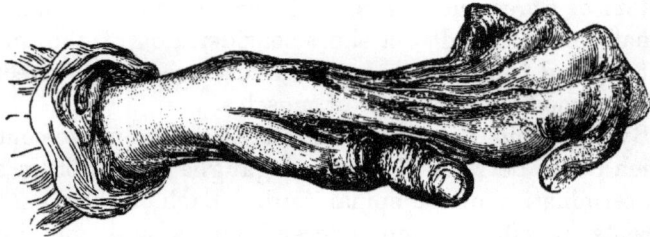

Fig. 12.—"Main en griffe" in a case of Progressive Muscular Atrophy.

insertion of a muscle remain approximated for a certain time, the muscle contracts and gradually shortens. This frequently occurs in joint disease, and after prolonged application of splints. In the case of progressive muscular atrophy, the "claw" condition results mainly from atrophy of the interossei. These muscles extend the two distal phalanges, and flex the proximal; and when their influence is removed, the opposite condition prevails, owing to the action of opponents, namely, extension or hyperextension of the proximal phalanges, and flexion of the peripheral.

The case from which the drawing was taken was kindly sent me by Mr. Sidney Harvey. The patient was a woman, aged 27, who was healthy in other respects, and the mother of a family. The disease had commenced in the left hand at the age of seventeen or eighteen.

Another instance of a similar kind is found in the disease called "pachymeningitis cervicalis," where the contraction varies according to the part of the cervical region which is affected. If that part from which the median and ulnar nerves emanate be diseased, the muscles supplied by the musculo-spiral nerve, remaining intact, produce the form of distortion in which the wrist is extended; while, if the upper half of the cervical region be the seat of the disease, the muscles innervated by the median and ulnar nerves produce quite a different form of distortion, in which the wrist is flexed.

Great interest has been taken of late in the effects of alcohol upon peripheral nerves, and some of our own countrymen have played a prominent part in the investigation of them—Wilks, Buzzard, Dreschfeld, Hadden, and others. There is very little doubt that a multiple peripheral neuritis results from the circulation of this poison, and produces such symptoms as neuralgic pains, anæsthesia, paræsthesiæ, various degrees of paralysis and atrophy, mainly occurring in the extensor muscles, ataxia, abolition of tendon-reflexes, vaso-motor disorders, and occasionally rigidity. Although our knowledge of this subject is not very advanced, it is probable that this rigidity is of the kind we are now considering, namely, rigidity of the more healthy muscles which overpower their diseased opponents; and, the latter being usually the extensors of the legs, rigid flexion results. Similarly, in cases of disease of the peripheral nerves and muscles on the extensor surface of the forearms produced by lead, rigidity of the opposing flexors is sometimes observed. As I have had an opportunity of making a post-mortem examination in a case of atrophy of the nerves and muscles of the forearm from lead-poisoning, I have brought drawings which illustrate peripheral neuro-muscular disease due to this cause.

The drawings were made from microscopic preparations of the muscles of a patient, aged 48, a plumber, who was admitted into St. Thomas's Hospital under the care of Dr. Bristowe, on March 18th, 1882, and died three days after admission from coma following an epileptiform convulsion. The patient had had several attacks of intestinal colic and of gout, and had also had " dropped hands " five or six years previously, from which he said he made a complete recovery. For the two months preceding his admission he had been labouring under general dropsy and wrist-drop. His arteries were found to be rigid, his

heart hypertrophied, and his urine albuminous, with
specific gravity 1011. He had paralysis, with marked
wasting of the muscles of the forearm, mainly the extensors, and also of the thenar and hypothenar muscles, and
of the interossei.

At the post-mortem examination, the heart was found
to weigh twenty-two ounces, the valves being competent.
The kidneys were contracted, hard, and granular on the
surface, and weighed eight ounces. The common extensors of both forearms were pale and atrophied, and quite
unlike the healthy muscles, the supinator longus being
among the latter. The posterior interosseous nerve was
soft and greyish, while the median and others were firm
and white. The microscopic sections of the extensor
muscles showed the following peculiarities.

Longitudinal section.—(1) Here and there were seen
well-formed, large, striated muscular fibres; (2) atrophied,
but striated, fibres, many extremely thin; (3) waxy-
looking, non-striated fibres, some of fair size, some very
small; (4) strands of wavy tissue where no muscular
fibres could be distinguished, but where long rows of
nuclei were seen, like necklaces, often running parallel to
each other, and quite close together; (5) very thick-walled
vessels.

Transverse section showed the muscular bundles very
small, each one having a homogeneous appearance, and
presenting nuclei irregularly distributed upon it. The
nerves were so altered as to be incapable of recognition.
The sections of the spinal cord in the cervical region
showed no evidences of disease. The supinator longus,
as is well seen in the drawing, was practically healthy.

In all these and similar cases, the diagnosis depends
upon the electric reactions of the nerves and muscles in
the affected part. For, where the rigidity is directly
produced by contraction of the affected muscles, the
electric conditions in them closely resemble those of
health; whereas, when the distortion is due to atrophy of
one set of muscles, owing to disease of the peripheral

nerves supplying them, what is called the "reaction of
degeneration" is found in them.

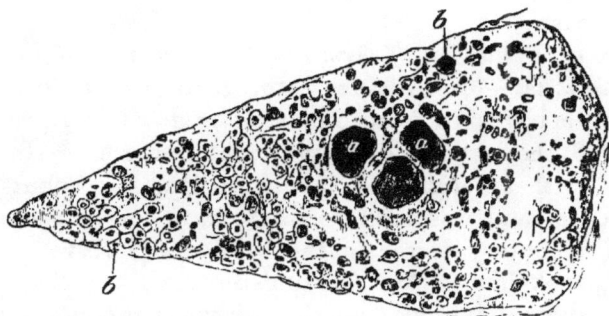

FIG. 13.—Transverse section of the Common Extensor of the Forearm
in a case of Lead-palsy. The darkly shaded masses in the centre (*a*)
are the largest muscular bundles left, most having been reduced to
very small dimensions (*b*).

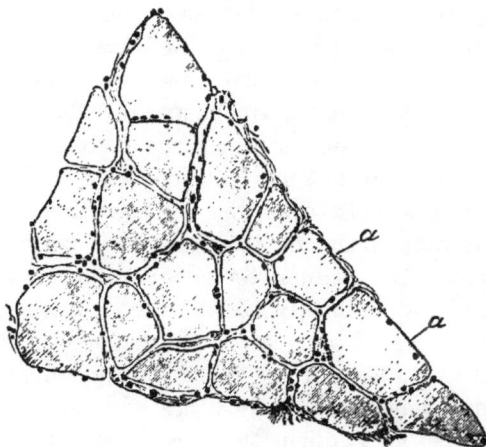

FIG. 14.—Transverse section of the Supinator Longus Muscle from a case
of Lead-palsy, showing the unaltered muscular bundles (*a a*).

Setting aside such cases as these, in which the distortion is produced by rigidity, not of the diseased muscles,
but of their healthy opponents, there exists still another

class in which contracture results from the shortening of
diseased muscles, it is true, but nevertheless does not depend
upon true spasm.   We may take as an example the following case :

*Case of Distortion of Arm due to Shortening of Degene-
rated Muscles.*—On August 11th, 1885, I saw, with Mr.
Clutton, a boy aged 5, who had fallen down in April and
fractured the lower end of the right humerus.   He was
treated with anterior and lateral rectangular splints for
one month, and was then found to have loss of sensation
and of voluntary power in the hand and forearm.   The
friends took him to Mr. Golding-Bird, who applied electricity daily for one month, without producing any
improvement.   When I saw the boy, his right forearm
was rigidly pronated, and the wrist partially flexed.   The
fingers were flexed at the peripheral joints, and slightly
over-extended at the metacarpo-phalangeal joints.   On
attempting to overcome this condition and extend the
wrist, the deep structures along the flexor surfaces of the
ulna became tense and hard, and, on bending up the
elbow-joint, they were again relaxed.   The tips of the
fingers were anæsthetic as far as the proximal phalangeal
joints.   The thumb had not lost sensation.   The question
which presented itself to us was whether this rigidity was
due to muscular spasm or to shortening of muscles, &c.,
from disease.   Dr. Kilner tested the muscles electrically
with me, and found that, while all the extensor muscles of
the forearm reacted normally, the flexors could not be
brought into action even by the strongest currents.   The
muscles were therefore clearly in an advanced condition of
degeneration.   To confirm this conclusion, the child was
put well under chloroform, but no change whatever
occurred in the limb ; all was as rigid as before.

Here, no doubt, we had to do with inflammation and
subsequent degeneration of nerve and muscle, and the
cicatricial contraction which accompanies them.   Electricity
and chloroform are our principal aids in diagnosis of such
cases.

Dr. Alliott, of Sevenoaks, kindly sent me a man, aged 65, who had dislocated his left shoulder-joint five months previously. On examination of the arm, almost complete paralysis of all the muscles was found, with but little alteration of sensation. The elbow, wrist, and fingers were all somewhat flexed, and incapable of extension; but still there was no active contraction of muscles. Even the strongest faradic current produced no effect whatever in them.

In certain cases of long-standing facial paralysis, the muscles on the affected side are evidently shortened, for the angle of the mouth is drawn to that side, and the orbicularis orbis partially closes the corresponding eye. And yet the effect produced by speaking or by laughing is most curious, the affected side remaining motionless. In these cases, the electric reactions of the muscles supplied by the diseased facial nerve show almost complete degeneration, which has been accompanied with permanent shortening.

If so many cases of distortion are due to other causes than active contraction of the muscles supplied by the affected nerves, the question suggests itself : Does chronic muscular spasm occur at all from direct irritation of motor nerves ? Let us take some of the most frequent and familiar examples of peripheral nerve disease. Musculo-spiral paralysis is often seen in quite the early stages ; but we do not see, nor do we hear the patient describe, muscular spasm as one of its symptoms. The same may be said of most cases of facial paralysis due to peripheral causes. In chronic poisoning by lead and alcohol, various nerve-symptoms are met with, including paralysis, and even muscular rigidity occasionally; but, as has already been pointed out, this is due to the contraction of antagonists, and not of the diseased muscles themselves. In sciatica, from various causes, we meet with twitchings and sudden muscular contractions; but they are very transitory. Again, if we look at those cases where motor or mixed nerves are pressed upon and irritated by new growths,

pain is a very prominent symptom, and spasm seems
scarcely to occur, though twitchings and transitory contractions are observed. A case which I have already
related affords a good example of a tumour pressing upon
a purely motor root. The new growth was a myxoma,
which grew in connection with one of the left anterior
spinal roots, and gave rise to symptoms in the left arm.
These were weakness and attacks from time to time of
pain and cramp; but no prolonged spasm was ever produced.

Dr. Bristowe has written a paper on Painful Paraplegia
in the ' St. Thomas's Hospital Reports,' vol. xii, in which
he relates several cases of paralysis accompanied by pain,
which owed their origin to malignant tumours involving
peripheral nerve-roots. Some of these cases I saw myself
and also examined post mortem. Evidence of spasm
occurs in only one case, where the limbs are said to have
become stiff occasionally.

Clonic spasm is not unfrequently seen in late stages of
facial paralysis, when there is shortening of the muscle
from atrophy. Electric examination shows that the latter
condition is present, and also that a certain amount of muscular tissue, capable of active contraction, remains. The
clonic spasms result from irritation of the nerve and muscle
which has survived the process of destruction. In cases
of meningitis at the base of the brain, spasm of the
muscles supplied by the motor nerves involved is not
observed. In spinal meningitis, on the contrary, it is a
very marked symptom. Here, however, it is not improba-
bly of a reflex origin, and not due to direct irritation of
motor nerves.

Tetany (Fig. 15) is a disease in which the presence of
some, although generally trivial, alteration of sensation,
increased electric irritability of nerve and muscle, and
prolonged muscular spasm, suggest a peripheral cause.
The pathology of this affection cannot be said to be
known, and opportunities of examining cases post mortem
rarely fall to the lot of pathologists. Dr. Hadden kindly

gave me the opportunity of examining microscopic sections
of the nerves and muscles of the affected parts, which
were taken from a child who died from diarrhœa, while
suffering from tetany. At first I thought there were
changes to be made out; but, on procuring sections from
the same parts, in a healthy child of the same age, 16

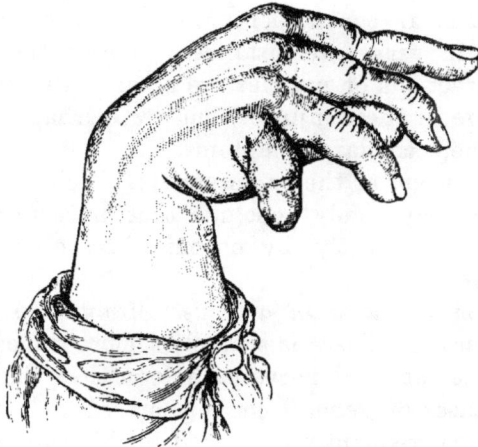

Fig. 15.—The Hand in a case of Tetany.

months, who had died of an acute disease, and, comparing
them with those from the subject of tetany, I came to the
conclusion that there was very little difference between
them.

Looking at my own experience, I should say that,
while occasional spasm occurs not unusually in peripheral
nerve disease, from direct irritation of the motor filaments
and may even be permanent in disease of the facial nerve,
it is quite the exception to find muscular contractions
among the marked phenomena of such cases.

Weir Mitchell, who has had a very large experience of
injuries and diseases of peripheral nerves, writes thus on
the subject ('Injuries of Nerves and their Consequences,'

1872) :—"Increase of bulk, proliferation of connective
tissue, and wasting of nerve-tubes, are consequences of
chronic neuritis." "The nerve-tubes in large part perish
or waste, and the symptoms affect at first rather the sensory sphere than that of motility. We have pain and
anæsthesia, or hyperæsthesia, but not, as a rule, local
convulsions." "In certain cases, the nerve-wound, in
place of causing primary loss of motility, occasions either
sudden muscular contraction, followed by instant loss of
power, or, in very rare instances, long-continued spasm.
Tonic contractions of muscles are occasionally met with at
a later stage of these injuries, but are, perhaps, among the
rarest of the secondary symptoms."

Looking, then, at the experience of Weir Mitchell and
others, we may safely conclude that muscular spasm is
rarely caused directly by chronic disease of peripheral
motor nerves.

2. *Spasm produced in a Reflex Manner by Disease of
Afferent Nerves.*—There is scarcely a more difficult subject
in the whole range of nerve-pathology than that of reflex
spasm. Cases of general convulsions, as well as of local
spasm, are so frequently explained in this way, and are in
so many instances quite insufficiently supported by scientific
evidence, that one cannot help coming to the conclusion
that many cases are accepted as of reflex origin which, at
any rate, carry very little conviction with them. Some
reasons can be adduced for the difficulties which beset the
diagnosis. Many of these cases occur where some sensory
nerve is the seat of severe pain ; or where a part is diseased
which is painful to move. A good example of the former
class is spasm of the muscles of the face, accompanying
facial neuralgia ; and, of the latter, rigidity in joint disease.
But the most ordinary method of expressing pain is by
some overaction of the facial muscles; while the ordinary
way of preventing pain in joint-disease is by keeping the
joint still, and opposing attempts at movement by contraction of muscles which produce an opposite effect. This is
so natural a contrivance, not only in man but in lower

animals, that it must be looked upon as almost an involuntary act.   Reflex muscular acts are in very different
degrees capable of being controlled or modified by the will ;
and even if it be allowed that the latter plays a part in the
limitation of the movements of a diseased joint, this does
not prove that the act is altogether involuntary and reflex.
In many cases, it is very difficult to form a definite conclusion with regard to this question.

No one who observes the great variety in the degree of
reflex muscular contraction produced by similar stimuli
applied to different individuals, can be surprised at finding
evidence that persistent spasm may sometimes occur as the
product of a reflex act ; or that a stimulus, which produces
no motor result in one person, gives rise to definite muscular contractions in another not equally healthy.   Thus,
in hemiplegia accompanied by descending sclerosis in the
lateral columns, deep reflexes are much more brisk than
they were before the hemiplegia occurred ; and contractures, Brissaud says, may be suddenly increased by comparatively slight injuries.   If this increased reflex excitability be due to the hyperphysiological activity of the
spinal centres, which have been loosed from cerebral control, similar disorders of nerve-centres rather than of
nerve-fibres are probably the most fruitful causes of reflex
spasm.   Hence, it is scarcely too much to say that the
injury, or disease, which supplies the stimulus to the sensory nerve in such cases, though apparently the principal
agent, is really so in many instances only from a particular
point of view.   That is to say, a mine exists in the patient
which has been accidentally exploded by a stimulus applied
to a nerve in connection with it ; but, had the same nerve
been connected with a healthy and stable centre, no spasm
would have ensued.

In speaking, therefore, of the present division of the
subject, namely, spasm produced in a reflex manner by
disease of afferent nerves, we can scarcely deal with it
apart from the next division, namely, spasm produced by
disease of ganglionic cells.   For it is questionable how far

stimuli applied to afferent nerves in chronic disease would produce muscular spasm if the centres were healthy. I am not aware that there is evidence to show that gross or demonstrable disease of ganglionic cells produces spasm. But impalpable disorders do; such, for instance, as those resulting from their separation from the cerebral centres. Non-demonstrable disorders, however, are what we presume to be the basis of so-called functional spasms, and to these we shall revert presently.

Reflex spasm no doubt occurs, but how frequently it does so, or how far the afferent or efferent nerves, or the nerve-centres, take the leading part in its production, are points which can scarcely be estimated. As good an instance as I know of reflex spasm is recorded by Mr. Clutton in the 'St. Thomas's Hospital Reports,' vol. x, p. 64. A boy, aged 14, had been bitten in the face by a dog eighteen months previously. The spot had been painless until a month before his appearance at the hospital; but, since that time, he had suffered from constant shooting pains in the neck, which always started from the scar. At the same time that the pain was felt, the angle of the mouth was drawn outwards, and the skin of the neck was wrinkled by the action of the platysma; the whole side of the face and neck blushed, and then became bathed in perspiration. This succession of symptoms recurred every time the scar was pinched. Croton-chloral-hydrate in five-grain doses twice a day soon cured the affection.

Weir Mitchell, in his work on 'Injuries of Nerves,' gives instances of reflex muscular spasm; and this condition is often referred to intestinal, uterine, or other irritation, but not always with sufficient reason.

Charcot has written a good deal in support of the view that various affections of joints may give rise to rigidity of muscles in a reflex way; and it certainly is very difficult to explain them otherwise. The following case appears to me to be one of the kind referred to.

Case of Persistent Muscular Spasm due to Joint Disease.—F. W—, æt. 16, was an out-patient at St. Thomas's

Hospital, under the care of Mr. Clutton, in February,
1884. I saw the case with him, and the following is the
clinical record which he took while she was under his
treatment, and which he has kindly put at my disposal.
The patient was sent to him by Mr. Merces, of the Bath
Mineral Water Hospital, with the following history.
During May, 1883, she had had rheumatic fever, and the
joints chiefly affected were the wrists and ankles. For
three months the wrists were placed in splints, and in
the autumn of the same year she was sent to Bath with
perfectly stiff fingers and wrists. Passive movement was
employed, which made the fingers pliable, but which left
the wrists as stiff as ever. Both hands were in the same
condition. On attempting to move the wrist-joint, the
flexors and extensors alike resisted, and could not be
overcome. Taking them by surprise produced no better
results. When the patient was placed under ether, the
muscles yielded to attempts at flexion and extension, but
the rigidity did not entirely disappear. There were but
few adhesions in the joint, and these readily gave way
under the anæsthetic, but left the state of affairs practically the same as before. That is to say, the wrist
could be flexed or extended, but, on removing the force,
the hand again assumed the same position. When the
wrist was extended, the fingers always became flexed,
and when it was flexed, they became extended. The
patient could voluntarily extend or flex the phalanges,
and could separate the fingers. The thenar muscles also
acted normally. The wrist was rigid, and could not be
moved in any direction, and there was slight hyperextension at the metacarpo-phalangeal joints, which was
easily overcome voluntarily when the patient flexed the
fingers. Dr. Kilner tested the electric condition of the
nerves and muscles, and found it normal.

Let us here briefly refer to a class of spasmodic affections which appear to have the same explanation as tendon-
reflexes. Whether explicable as a reflex or as a purely
local phenomenon, contraction of muscles occurs when they

6

are put on the stretch by their antagonists ; and this has
the effect of controlling and steadying movements which
might otherwise be jerky and uncertain. When the
influence of the lateral columns of the cord is removed, the
spinal centres, in which reflex acts are produced in relation
with parts below the level of the lesion, are brought into
a state of excessive activity, and stretching of a tendon
produces an abnormally sudden and extensive contraction.
This is very probably the cause of the tremors in the
direction of movement seen in cases of disseminated
sclerosis. But it is probable that some cases may have a
peripheral origin. A medical friend of mine informed me
that he once took a vehicle, in order to drive to a house
where he was going to stay for a while. The driver put·
him down two and a half miles from his destination, and
drove off. The medical man had rather a heavy bag, and,
being unable to get another conveyance, he carried it
himself. For four days after this walk he suffered from
pretty constant contractions of the triceps muscle of his
right arm whenever he flexed it. In this instance,
the muscle had been what we call strained, the tendon had
been unduly pulled upon, and, when the arm was flexed,
the slight increase of tension made the triceps contract.
Such a case represents a very small departure from the
normal, but suggests an explanation of more troublesome
affections ; such a one, for instance, as the following.

*Case of Persistent Rhythmical Contraction of the Pal-
maris Longus, due to Injury.*—On July 7th, 1878, I saw a
girl, aged 19, who had had her wrist bent backwards and
sprained five years previously. The parts injured had
swelled, and she had been obliged to carry her arm in a
sling for some weeks. Ever since that time she had
suffered without intermission from twitchings of the arm
and palm of the hand. On examination, I found that the
general power of the arm was unimpaired, but that there
were spasmodic contractions of the palmaris longus muscle,
occurring with perfect regularity ninety times in the
minute. The contractions were energetic, and showed the

muscle and its attachments very beautifully. In this case probably the palmaris longus had been stretched, and the peripheral nerves ending in it had been rendered more irritable, the result being that the slight tension of it which occurs ordinarily in the act of extending the wrist, caused the muscle to contract. This occurring frequently, gave rise to a neuro-muscular habit.

The following is a somewhat similar instance. Mr. Clutton once showed me a girl, aged 10, who had fallen down a week previously and slightly sprained her right wrist, but had done herself no other damage. I found her suffering from an affection which had come on immediately after the fall. Her right arm was in a position midway between pronation and supination, and was the seat of perfectly regular tremors of very small excursion, such as one sees in paralysis agitans, occurring in a direction transverse to the longitudinal axis of the limb.

Other pathological conditions, as well as injuries, may produce these rhythmical spasms. In January, 1886, a girl, aged 14, was in St. Thomas's Hospital, under the care of Dr. Bristowe, for rhythmical contraction of the occipito-frontalis muscle on both sides, which occurred from fifty to sixty times in the minute, and had been going on for some months. The history she gave was that she had had a very bad attack of " erysipelas of the face," commencing on the forehead and spreading downwards. The onset was sudden, and the description she gave of the affection corresponded with that of erysipelas. Roughness and pigmentation of the skin of the forehead could still be seen. As the disease got well, the contraction of the occipito-frontalis muscle supervened, and continued for months without intermission, except during sleep. The original cause of this condition was probably an irritable state of the muscle and tendon owing to inflammation, and an involuntary neuro-muscular habit was soon developed. Although the girl showed no evident hysterical peculiarities I found that she had marked localised tenderness over those regions which Charcot has found to be unduly

sensitive in hysterical subjects.   Moreover, information
which I subsequently obtained proved that at least one
other member of her family was hysterical.   For, while
the patient was still in the hospital, her sister, aged 11, was
admitted under Dr. Bristowe's care, suffering from a similar
affection.   The latter had been under me as an out-patient
for fits, which were evidently hysterical ; and, at the early
age of eight years, she had had an attack of rhythmical
spasm of the occipito-frontalis muscle which lasted three
months, and then got well.   When I saw her in the hospital during the second attack, I found that the muscle
was contracting pretty regularly 120 times per minute,
and this condition was accompanied with sighing, yawn-
ing, slight movements of the tongue, partial loss of sensation on the left side, pain in the left hip and knee, and
drawing up of the left leg so as to produce distinct
shortening.

Other cases of spasm from peripheral injury are more
complicated, but appear to be reflex in their origin, whether
the pathological stimulus acts upon the nerves which exist
in muscles and tendons or upon other afferent paths.   The
following is an example.

*Case of Spasmodic Movements of the Jaw, the Floor of the
Mouth, &c., due to Injury.*—On November 20th, 1877, a
boy, aged 10, came to St. Thomas's Hospital, on account of
a peculiar affection which he had had for five days.   There
was nothing noteworthy in his family or personal history,
except that he had suffered from incontinence of urine.
Five days before coming to St. Thomas's he had been in
a playground where his schoolfellows were having a
game of football.   He happened to be in their way, and
one of them took him by the back of the neck, and threw
him to one side.   His neck pained him at the time, but
by the following day the pain was gone.   In its place,
however, was a peculiar affection, consisting of constant
movements of the lower jaw and of the floor of the mouth,
and they had increased since then.   As he stood in front
of me, I saw the lower jaw and the floor of the mouth

descend in regular and frequently recurring jerks, and when
the child's mouth was open it was seen that the tongue
was likewise jerked spasmodically downwards and backwards.   In addition to this, the larynx descended slightly,
and the depressor muscles in the neck could be seen contracting.   Each downward movement was accompanied by
a sound like the croaking of a frog.

Neither the faradic nor the galvanic current applied
from the nape of the neck to the parts below the jaw had
the slightest effect on this curious condition.   I gave the
child some bromide of potassium, and sent him away.   Nine
months after that I saw him, and he was nearly, but not
quite well.

As I have already said, it is a matter for speculation how
far such causes would give rise to spasm, if the nerve-centres
were in a normal condition.   What is really developed
in these cases is an involuntary neuro-muscular habit.
We are all familiar with the ankle-clonus which is set up
sometimes in healthy people when sitting with the toes on
the ground and the heels a little raised.   If we start the
clonus voluntarily, or if it has occurred several times in
succession involuntarily, it may be difficult to stop it by an
effort of the will, unless the position of the legs be altered.
This is probably a physiological representative of certain
cases of muscular spasm which assume pathological
dimensions.

I have now come to the end of the remarks I have to
make upon the first two divisions of spasmodic affections
in connection with chronic nerve disease, those, namely,
which result from pathological changes in the cerebral
motor mechanisms, on the one hand, and the spinal motor
mechanisms on the other.   The third division, and that
the most difficult of all, remains still to be dealt with, and
is co-extensive with what are called functional spasms.
As I have already pointed out, the first two divisions
should, and we may hope some day will, cover the whole
field of spasmodic affections ; and, by placing functional
diseases under a separate heading, we publicly confess our

ignorance and proclaim that there is a very long list of
spasmodic affections which have been carefully observed
without success, so far as their pathology is concerned. In
lectures professing to deal, however perfunctorily, with
spasm, it would be impossible to pass by in silence the
class referred to. It remains, therefore, for me to consider how far these functional diseases are capable of
explanation or reasonable classification.

III. FUNCTIONAL SPASMS.

The result of our consideration of those diseases which
produce many varieties of spasm, and which are represented
by gross lesions of the nervous system, is, that a very large
proportion must be put down to the credit of the cerebral
system of fibres, or pyramidal tract. The spinal mechanisms are less prolific causes of these affections ; but those
spasmodic conditions which do occur in connection with
them have been considered under the heads : (1) peripheral
motor nerves ; (2) afferent nerves ; (3) centres. It is only
reasonable to suppose that, if functional spasms resemble
those which result from gross lesions, they may admit of a
similar explanation. In nerves and muscles which during
life have been the seat of motor disorders, we may find no
naked-eye or microscopic pathological changes after death.
But clinical observation shows that, so far as function is
concerned, marked changes have occurred. Now, we know
that healthy function must depend upon fine molecular
rearrangements, which are at present a closed book to us ;
and we conclude that, if molecular changes form the basis
of healthy function, they form that of pathological conditions also. If this be so, it is legitimate to argue that the
molecular changes in question might affect the same parts
which are the usual seats of the grosser lesions, and consequently give rise to similar abnormalities of function ;
and then the anatomical classification which has already
been attempted in the case of tangible disease, might serve

also for those cases which are the result of impalpable alterations in nerve and muscle. I propose, therefore, to consider shortly certain forms of functional spasm, and I shall use the same divisions for classification which I have already used for grosser diseases.

1. Do functional disorders of the cerebral motor mechanism, or pyramidal tract, produce spasmodic conditions at all resembling those resulting from gross disease? They do. In hysteria, not only does hemiplegia occur, but hemispasm, both fixed and mobile, is not uncommon. There is this difference between the hysterical conditions and those which are seen in gross disease; the leg is affected most, the arm less, and the face not at all. But the facts observed in gross disease, where the leg is usually less affected than the arm, are probably to be explained mainly by the fact that the vessels which bleed most frequently, and interrupt the functions of the internal capsule, are those which are situated towards the anterior part of its motor division, that part, in fact, which contains the fibres controlling the face and arm. But apart from these points, the spasmodic contractions of the leg and arm are similar in the two conditions. In functional diseases, there is simply removal of the voluntary impulses which, in health, travel down the pyramidal tract. The muscles and nerves of the limbs remain practically healthy; and if rigidity be not too great, the tendon-reflexes are increased. Fevers and other exhausting conditions, such, for instance, as that of the patient with anæmia to whom I have already referred, may give rise to increased reflexes, and even to clonus and rigidity. Moreover, absence of portions of the motor area of the cortex, such as is occasionally seen congenitally, is accompanied by non-development of the corresponding portion of the pyramidal tract, and gives rise to spastic conditions like those found in hemiplegia. Absence of the voluntary impulses, then, is enough to give rise to the phenomena in question; and certain cases of hysteria have died after long-continued contraction of limbs, and the most careful examination with the microscope has failed

to disclose any palpable pathological changes. Such a case is recorded by Dr. Bristowe in ' Brain ' for October, 1885. This case I often saw myself, and, as Dr. Bristowe remarks, " at the end of her two years, she seemed as well in general health as when she first came to the hospital ; but she was suffering from headache, sickness, ophthalmoplegia externa, complete anæsthesia of the right side, with rigid paralysis of the arm and leg, and repeated hæmorrhages of both ears." Dr. Hadden made a complete microscopical examination of the central nervous system in this case, but found no evidences of disease. As we know, however, that mere absence of the voluntary motor impulses along the pyramidal tract is sufficient to give rise to this condition, it is legitimate to conclude that suppression of the functions of this tract is the cause of functional hemispasm. We are still further supported in this view by cases like that just quoted, where there is at the same time clear evidence of the suppression of the functions of the sensory area as well.

Clinical and pathological observation has shown that monospasms having similar characters may be produced by certain limited lesions of the motor convolutions ; and, after what has already been said, it is not too much to assert that when spasm limited to a leg or arm occurs in hysteria it is due to suppression of the functions of the pyramidal tract, which in health controls the action of the affected member. In discussing the question of athetosis and other allied mobile spasms, we arrived at the conclusion that they were due to affections of the same tract of fibres which were not severe enough to stop the passage of voluntary impulses. Now, similar conditions occur in hysteria, and may probably be explained in the same way. Weir Mitchell, in his work on ' Nervous Diseases,' records cases of " hysterical motor ataxy " and of " hysterical athetosis." It appears probable that changes, ending in greater or less functional abeyance of the pyramidal tract, lie at the root of unilateral, and sometimes of bilateral spasm, both in functional and in gross disease. The only difference is that, in the latter case, we can appreciate, by our present methods of investigation, the alterations which have been wrought, while in the former, they are hidden from our view.

There is another class of functional spasmodic affections which are hardly represented among the forms of spasm due to gross lesions, and which, I believe, have their origin in suppression of the functions of certain portions of the pyramidal tract. I refer to those conditions which are sometimes termed " professional hyperkineses." They include histrionic spasm, writers' cramp; pianists' cramp, telegraphists' cramp, &c. The striking peculiarity of these affections is that, when a voluntary effort is made to perform the accustomed muscular acts, spasm of the muscles involved occurs and prevents further efforts. Before this stage of the affection is reached, great fatigue often accompanies endeavours to work. In a considerable number of cases, at any rate, the muscles can be employed in other ways without difficulty. What has occurred is probably that, after long-repeated acts of the same kind, that part of the pyramidal tract which is used becomes fatigued and its functions are partly suppressed, so that a condition of " latent contracture" of the muscles, over the voluntary actions of which it presides, is developed ; the lower centres are " let go," as Hughlings Jackson says, and are in a state of " hyperphysiological activity," just as the spinal centres of one side are in hemiplegia accompanied by slight rigidity.

In describing " latent contracture " in connection with hemiplegia from gross disease, Ross says : " The patient may perform all the simple movements of the limb, and probably with undiminished power ; but when his attention is specially directed to the movements, as when he wishes to perform any manual operation requiring a little dexterity, the muscles instantly become rigid, the fingers are flexed on the palm, and the deformity, which was present during the period of fixed contracture reappears." Surely this description suggests the true explanation of professional

spasmodic affections, which present symptoms resembling
very strikingly those which Ross portrays.

Erb divides the professional hyperkineses into spastic,
tremulous, and paralytic. The tremulous cases are perhaps
due to similar but slighter paralytic affections of the pyramidal tract, and the condition probably resembles in kind,
though not in degree, paralysis agitans. The pathology
of the latter disease is not known, but those cases which
one meets with where rigidity of limbs and exaggerated
reflexes occur, suggest forcibly degeneration in the pyramidal tract as the physical basis of the affection. The
paralytic variety of the professional hyperkineses is probably the result of extension of disease to the motor cells
in the anterior cornua of the spinal cord, at any rate in
those cases where marked atrophy occurs (see ' Brain,'
vol. vi, 233, a " A Case of Sawyers' Cramp," by G. V.
Poore, M.D.).

I should remind my hearers, in passing, that the term
pyramidal tract, as used here, includes not only fibres, but
the motor cells of the cortex in which they originate.

2. We have seen that spasms of muscles resulting from
gross disease of the pyramidal tract has its representative
among functional diseases; let us now consider that which
owes its origin to affections of peripheral motor nerves.
Is it likewise represented among functional disorders?
Actual gross pathological changes in motor nerves give
rise to muscular spasm, as has already been pointed out,
in two ways—the one direct, and the other indirect. In
the latter case, certain muscles atrophy and disappear, and
undergo cicatricial shortening, producing distortion; or
else their opponents being left to act in their absence, give
rise to abnormal positions of the limbs. Now, one of the
striking peculiarities in functional nerve disease is the
absence of such degeneration and destruction of muscles.
Hence we should not expect to find any functional spasms
due to such indirect causes; nor do we, so far as I know.
We also came to the conclusion that direct irritation of
motor nerves is a very rare originator of spasm; and there

is no reason to suppose that it is less rare as a cause of functional affections. Nerve-centres seem much more prone to functional disorder than nerve-fibres. I am not sure that the latter condition, originating in centres, may not so affect even peripheral nerves which are connected with them, as to give rise to symptoms which are found in gross disease only when peripheral nerves are involved. Dr. Norris, of Windsor, sent to me, for treatment in the hospital, a girl, aged 16, who had lost the use of her left hand for eighteen months. She was a strong, healthy girl and free from evident hysterical tendencies. The affection commenced with swelling and blueness of the fingers, such as is seen in chilblains ; and the hand was cold and numb. When she came to me, the hand was completely paralysed, and had a swollen puffy look ; the skin was paler and smoother than that of the right hand, and the backs of the fingers far less wrinkled. The left hand was much more influenced by external conditions than the right, and rapidly became warm when covered, and cold when exposed. She suddenly recovered after the application of a blister to the wrist, but then lost power in the left leg. The latter also suddenly recovered power. But, although slight alterations, such as those which were observed in this patient, may occasionally occur in peripheral motor nerves in cases of functional disease, I am not aware that they give rise to muscular contraction.

3. Nor is there any proof that chronic spasm is produced by molecular changes of afferent nerves alone. Even in gross disease of an afferent nerve giving rise to reflex spasm, the lesion is often so trivial as to make it very probable that functional disorder of the centres is present as well, and is the main factor in the production of spasm.

When speaking of reflex spasm from gross disease, I pointed out certain cases of what appeared to be rhythmical spasm due to affections of tendons ; and I cannot help thinking that this class has frequent representatives among functional diseases. But they are due more to the

abnormal condition of the centres than to that of the peripheral nerves.   There can be no question that one of the
peculiarities of nerve-centres in hysteria is their abnormal
irritability ; so that slight afferent impressions give rise to
muscular acts, which pass with great ease into neuro-
muscular habits.   Such a condition is well exemplified in
cases due to imitation.   The patient sees a certain form of
muscular spasm, and the idea produces the same in her.
For instance, a healthy girl, aged 17, came to my out-patient
room complaining of involuntary movements in her right
hand and tongue.   On examination, it was seen that the
thumb was rhythmically adducted, and there was simul-
taneous flexion of the fingers, most markedly of the index,
which moved in such a way that she appeared to be rolling
something between it and the thumb.   At the same time,
she was troubled by being suddenly unable to finish a
sentence, owing to the tongue refusing to act properly.
The girl worked in a confectioner's shop, where one of her
comrades became affected with uncontrollable movements
in the left hand and catching in her speech.   After a short
time my patient became similarly affected.

The following is another case of suddenly developed
neuro-muscular habits, though it originated in a somewhat
different way. A healthy, dark-haired girl, aged 16, who had
never had any illness, and had no evident neurotic peculiarities, went out for a row on the Thames with some friends
in the summer of 1885.   She rowed without much intermission for five hours, which was to her quite an unaccustomed
length of time for such recreation.   She used to row every
morning at Wandsworth, where she was at school, but only
for an hour at a time.   About an hour after returning
from this long row, her arms began to twitch, and movements continued uninterruptedly, and remained of a precisely similar kind up to December 29th, 1885, when Dr.
Bristowe, under whose treatment she had been, very kindly
asked me if I should like to see her.

Both arms were the seat of similar and synchronous
movements, occurring regularly about one hundred and

sixty times a minute. They consisted of rapid elevation and retraction of the arm at the shoulder and of the scapula, partial flexion at the elbow, and slight extension of the wrist ; in fact, they bore a very marked resemblance to the movements of the arms and hands in rowing, the extreme rapidity of the "stroke" making up for the limited extent of the movements.

In these rhythmical contractions it is not at all improbable that the slight stretching of the tendons of the affected muscles which naturally results from the action of their antagonists gives rise to another contraction, and so on, for an indefinite number of times. Such a condition, though much less marked, is sometimes seen when the spinal centres are cut off from connection with the cerebral cortex by gross disease.

A case of caries of the cervical vertebræ was under my care in St. Thomas's Hospital last summer, in which paraplegia, with rigidity, had been present for five years. The legs were always flexed at the knee and hip, but could be extended by continuous, gentle traction. As this was being performed, movements of flexion and extension, which were involuntary, occurred. When the tendons of the flexor muscles were stretched, the latter contracted, and, in so doing, they put the extensor tendons on the stretch, and so caused their contraction.

Weir Mitchell has noticed these "alternate spasms" in hysteria, in which, as he remarks, the action of the flexors calls the extensors instantly into play, and this, in turn, summons the flexors into activity. "These spasmodic motions," says he, "were the more curious in the last case I saw, because of the general and profound paresis, which made every volitional effort excessively difficult." Such an absence of voluntary power, in the case of which Weir Mitchell speaks, shows that the functions of the cortical motor centres were in abeyance. Instances of alternating spasms are far from rare in hysteria, and probably owe their presence to excessive irritability of the spinal centres, and the development of a habit through the intermediation

of tendon-reflexes. " Saltatory spasm " is a good instance
of this class of cases.

4. The fourth and final heading, under which we have
to compare spasmodic conditions arising in connection with
gross nerve-lesions, with functional disorders, is disease of
the spinal motor centres. We have already seen that
tangible pathological alterations in these cells give rise to
paralysis and subsequent atrophy of muscles, but we could
quote no facts to show that muscular spasm ever originates
from such disease. Those impalpable molecular changes,
which result from the anatomical separation of the cerebral motor centres from the spinal centres, produce a condition of hyperphysiological activity of the latter, to which
many spastic conditions are due. Does abnormal functional activity ever exist primarily in the spinal centres
and result in spasm ? This is a question which is very
difficult to answer. In the case of gross lesions which
interrupt the continuity of the fibres of the pyramidal
tract, even where no actual spasm has resulted, there
exists a condition of excessive excitability of the spinal
centres, which gives rise, on the slightest provocation, to
rigidity of muscles. An injury, even of a trivial nature,
may, in such patients, produce contracture. Exactly the
same state of things is found in hysterical subjects ; and
Charcot, in his lectures on nerve disease, gives instances
of the most trifling injuries suddenly producing spasmodic
conditions of limbs in such patients. But we must
remember that persons may have those peculiarities of
their nervous system which are usually embraced under
the term hysteria, without ever presenting striking
emotional or other tendencies which are wont to call
attention to the existence of the disorder. The first
evidence of the latter may be the sudden supervention of
spasm from a very slight external injury. If this condition of the spinal centres depends mainly upon the amount
of cerebral control exerted over them by the functional
activity of the pyramidal tract, it is evident that there
must be infinite gradations between the normal and

abnormal. Indeed, it is impossible to give any strict definition of health and disease. All we can say is, that the more cerebral control retires into the background, the more likely are spasmodic conditions to come to the fore.

It is quite impossible to answer the question, whether hyperexcitability of the spinal centres occurs in the absence of diminution of cerebral influence to account for it ; but it probably does. We know that strychnine, in certain doses, produces an excitable condition of the spinal centres, which Charcot has likened to that which is seen in " latent contracture " due to gross lesions of the pyramidal tract, and to hysteria. In comparing strychninism with the latter, Weir Mitchell says (' Lectures on Diseases of the Nervous System, especially in Women,' page 100) : —" Perhaps I shall, in a measure, clear your minds as to the nature of what I mean by functional spasms if I recall to you the influence of strychnine in large doses, such as you have seen given here many times. You will remem- that in certain spinal maladies, such as those of syphilitic birth, it is my habit first to give iodide in heavy doses, then to suspend them for a time, and to give strychnine up to the limit of physiological endurance, that is to say, until I cause an approach to spasm. When, for example, you give hypodermically the one fifth to the one eighth of a grain daily, the patient will have little or no annoyance, if you are careful to insist that he remain at absolute rest in bed for two hours after each injection. If there be any tendency to spastic twitchings of the muscles, the will is competent to control them, unless—and this is the point I would make—the patient attempts to exercise. Should he do this the effort results at once in irregular move- ments of an inco-ordinate character, and in slight or more grave spasms of the muscles employed. While at rest there is no obvious trouble, but voluntary movement occa- sions spasms, which are the offspring of the poison. They are, in a word, functional spasms, and would not be seen at all with limited use of strychnine, were it not for the efforts of voluntary action." If, then, a substance circu-

lating in the blood can directly produce this condition of
the spinal centres, in which they may almost be said to be
watching their opportunity to produce muscular contraction, it is not improbable that molecular alterations may
arise from other causes, and produce a similar excitable
condition of these centres.   We know, however, little, if
anything, about the changes which act thus.   What are
called phantom tumours of the abdomen, and sometimes
of other parts, may have some such origin ; Weir Mitchell
records one, in which all the abdominal muscles had
remained violently contracted for a year.   But whether
this be the explanation of such cases or no, the hyperphysiological activity, due to lack of cerebral control, which
exists in the spinal centres of some subjects, is probably
the principal factor in a very large number of tonic and
clonic functional spasms.

In the comparison which has been attempted between
the results of gross disease on the one hand, and molecular or so-called functional alterations on the other, the
pyramidal tract stands prominently forth as the great
offender in the production of muscular spasm.   Its action
is indirect, it is true, as it only looses the reins of the
spinal centres, which it should keep well in hand.   Still,
it rules the situation.   Spasm rarely ensues directly from
injury or disease of peripheral motor nerves; and,
although it frequently results from reflex causes, it is
very likely that the spinal centres are, in a considerable
proportion of cases, more at fault than the afferent nerves.
It may be thought somewhat curious that, in speaking of
dimunition of cerebral control, the voluntary motor tract
should alone have been referred to as liable to suffer from
depression of its functions.   But the only reason why it
has been thus signalled out is, that these lectures have
dealt with motor phenomena.   The sensory functions of
the brain suffer, too, in a very striking manner, so that
complete hemianæsthesia is far from a rare occurrence in
hysteria, the loss of function being evident in the realm
of the special senses, as well as in that which has to do

with common sensation. And although at first sight these functional defects appear to be confined to one side, careful investigations show that they are not. For, in cases of hemianæsthesia, both fields of vision are contracted, although that on the anæsthetic side is the more affected. Again, in cases where the loss of function appears to be confined to the sensory area, a careful comparison of the muscular power on the two sides may show that this is not so. In a case of the kind which I lately saw, I found that on the hemianæsthetic side the patient could only reach 45 on the dynamometer scale, although this was the right side, while with the left she reached 50. In such hysterical patients, it is probable that nerve power is deficient in all parts and on both sides of the brain, although this deficiency is more marked on one side than the other, and in some portions of that side than in others. Considering how late the pyramidal tract develops in man, and what a high pitch of evolution it represents, it is not to be wondered at if it is one of the first parts to suffer in the process of dissolution.

If we are right in considering deficient nerve-power, especially in the brain, as the condition which is at the root of functional nerve disease, the term neurasthenia, a product of modern times, would be a more suitable general title for such disorders than the older term, hysteria. When we review the history of nerve disease in families, we cannot but conclude that people are born with very different physical bases, both in their nervous and in their other systems. And no line can be drawn between those who are normally stable and enduring under unfavorable circumstances, and those who are liable to lose equilibrium. Some patients would remain perfectly well had they not accidentally met with some catastrophe; and then they fall into the category of hysterical patients. And yet there are many of their friends and acquaintances who are as liable to functional disorder as they are, and who yet may never have their weakness brought to light, owing to their not having been exposed to a sufficiently severe trial.

7

Even small portions of nerve-centres may become hysterical, if I may be allowed to say so, as, for instance, in writers' cramp, pianists' cramp, and other professional hyperkineses. All that we know of this matter is, that functional nerve disease, even in the case of apparent exceptions, such as spasm, is due to deficiency of nerve-power. How the condition is actually brought about we cannot tell. Some think the centres are starved by contraction of their nutrient vessels. It seems to me more likely that the nerve-centres are exhausted by their own action. Thus, sudden emotional excitement may bring on hysterical phenomena, indicating nerve exhaustion, such as hemianæsthesia, paralysis, or spasm. It is important to remember, too, that exhaustion of one area seems to affect the whole system. This may indicate that the various differentiated centres draw nerve-force from some common supply. Considering the enormous amount of cells which are found in the cerebral grey matter, many without evident connections, is it going too far to suppose that a large number of these are factories for the production of nervous energy, which is necessary for the continuous and regular action of the highest centres, and which flows in increasing quantities towards those centres which are from time to time brought into action ? Constant supply would mean constant power in all centres ; and as the latter are all bound together so as to produce a physiological equilibrium, constant power would involve constant control of one by the other. Where the store of constantly produced force (represented, of course, by structural peculiarities) is by inheritance small, or where it has been exhausted by unnatural calls upon it, there would be insufficient to supply all the centres ; and those to which the nerve-paths are most freely open by habit would command it, while others would be starved, and loss of equilibrium, or the natural control of one centre by the healthy action of others, would be the inevitable result. If we do not suppose any such reservoir of nerve-energy, how can we explain the exhaustion of all, by the excessive action of some, centres ? Why is a person who has had excessive mental work less able for the time to do active bodily work ? and why, when bodily fatigued, are we less ready to do intellectual work ?

But I have gone, I fear, too far into a region of speculation, which it is hardly legitimate for me to have even entered with so small an equipment of facts. I shall therefore say no more, but bring these lectures to a conclusion by offering an apology for having chosen a subject which is so extensive in range, and at the same time so obscure and little understood, that my treatment of it must fall far short of that which is expected on occasions like the present.

PRINTED BY J. E. ADLARD, BARTHOLOMEW CLOSE

SELECTION

FROM

J. & A. CHURCHILL'S GENERAL CATALOGUE

COMPRISING

ALL RECENT WORKS PUBLISHED BY THEM

ON THE

ART AND SCIENCE OF MEDICINE

N.B.—As far as possible, this List is arranged in the order in which medical study is usually pursued

J. & A. CHURCHILL publish for the following Institutions and Public Bodies :—

ROYAL COLLEGE OF SURGEONS.
CATALOGUES' OF THE MUSEUM.
Twenty-three separate Catalogues (List and Prices can be obtained of J. & A. CHURCHILL).

GUY'S HOSPITAL.
REPORTS BY THE MEDICAL AND SURGICAL STAFF.
Vol. XXVII., Third Series (1884). 7s. 6d.
FORMULÆ USED IN THE HOSPITAL IN ADDITION TO THOSE IN THE B.P. 1s. 6d.

LONDON HOSPITAL.
PHARMACOPŒIA OF THE HOSPITAL. 3s.
CLINICAL LECTURES AND REPORTS BY THE MEDICAL AND SURGICAL STAFF. Vols. I. to IV. 7s. 6d. each.

ST. BARTHOLOMEW'S HOSPITAL.
CATALOGUE OF THE ANATOMICAL AND PATHOLOGICAL MUSEUM. Vol. I.—Pathology. 15s. Vol. II.—Teratology, Anatomy and Physiology, Botany. 7s. 6d.

ST. GEORGE'S HOSPITAL.
REPORTS BY THE MEDICAL AND SURGICAL STAFF.
The last Volume (X.) was issued in 1880. Price 7s. 6d.
CATALOGUE OF THE PATHOLOGICAL MUSEUM. 15s.
SUPPLEMENTARY CATALOGUE (1882). 5s.

ST. THOMAS'S HOSPITAL.
REPORTS BY THE MEDICAL AND SURGICAL STAFF.
Annually. Vol. XI., New Series (1882). 7s. 6d.

MIDDLESEX HOSPITAL.
CATALOGUE OF THE PATHOLOGICAL MUSEUM. 12s.

ROYAL LONDON OPHTHALMIC HOSPITAL.
REPORTS BY THE MEDICAL AND SURGICAL STAFF.
Occasionally. Vol. X., Part III. (August, 1882). 5s.

OPHTHALMOLOGICAL SOCIETY OF THE UNITED KINGDOM.
TRANSACTIONS.
Vol. V. (1884-5). 12s. 6d.

MEDICO-PSYCHOLOGICAL ASSOCIATION.
JOURNAL OF MENTAL SCIENCE.
Quarterly. 3s. 6d. each, or 14s. per annum.

PHARMACEUTICAL SOCIETY OF GREAT BRITAIN.
PHARMACEUTICAL JOURNAL AND TRANSACTIONS.
Every Saturday. 4d. each, or 20s. per annum, post free.

BRITISH PHARMACEUTICAL CONFERENCE.
YEAR BOOK OF PHARMACY.
In December. 10s.

BRITISH DENTAL ASSOCIATION.
JOURNAL OF THE ASSOCIATION AND MONTHLY REVIEW OF DENTAL SURGERY.
On the 15th of each Month. 6d. each, or 7s. per annum, post free.

A SELECTION

FROM

J. & A. CHURCHILL'S GENERAL CATALOGUE,

COMPRISING

ALL RECENT WORKS PUBLISHED BY THEM ON THE ART AND SCIENCE OF MEDICINE.

N.B.—*J. & A. Churchill's Descriptive List of Works on Chemistry, Materia Medica, Pharmacy, Botany, Photography, Zoology, the Microscope, and other Branches of Science, can be had on application.*

Practical Anatomy :
A Manual of Dissections. By CHRISTOPHER HEATH, Surgeon to University College Hospital. Sixth Edition. Revised by RICKMAN J. GODLEE, M.S. Lond., F.R.C.S., Demonstrator of Anatomy in University College, and Assistant Surgeon to the Hospital. Crown 8vo, with 24 Coloured Plates and 274 Engravings, 15s.

Wilson's Anatomist's Vade-Mecum. Tenth Edition. By GEORGE BUCHANAN, Professor of Clinical Surgery in the University of Glasgow ; and HENRY E. CLARK, M.R.C.S., Lecturer on Anatomy at the Glasgow Royal Infirmary School of Medicine. Crown 8vo, with 450 Engravings (including 26 Coloured Plates), 18s.

Braune's Atlas of Topographical Anatomy, after Plane Sections of Frozen Bodies. Translated by EDWARD BELLAMY, Surgeon to, and Lecturer on Anatomy, &c., at, Charing Cross Hospital. Large Imp. 8vo, with 34 Photolithographic Plates and 46 Woodcuts, 40s.

An Atlas of Human Anatomy. By RICKMAN J. GODLEE, M.S., F.R.C.S., Assistant Surgeon and Senior Demonstrator of Anatomy, University College Hospital. With 48 Imp. 4to Plates (112 figures), and a volume of Explanatory Text, 8vo, £4 14s. 6d.

Surgical Anatomy :
A series of Dissections, illustrating the Principal Regions of the Human Body. By JOSEPH MACLISE. Second Edition. 52 folio Plates and Text. £3 12s.

Medical Anatomy.
By FRANCIS SIBSON, M.D., F.R.C.P , F.R.S. Imp. folio, with 21 Coloured Plates, 42s.

Anatomy of the Joints of Man.
By HENRY MORRIS, Surgeon to, and Lecturer on Anatomy and Practical Surgery at, the Middlesex Hospital. 8vo, with 44 Lithographic Plates (several being coloured) and 13 Wood Engravings, 16s.

Manual of the Dissection of the Human Body. By LUTHER HOLDEN, Consulting Surgeon to St. Bartholomew's Hospital. Edited by JOHN LANGTON, F.R.C.S., Surgeon to, and Lecturer on Anatomy at, St. Bartholomew's Hospital. Fifth Edition. 8vo, with 208 Engravings. 20s.

By the same Author.

Human Osteology.
Sixth Edition, edited by the Author and JAMES SHUTER, F.R.C.S., M.A., M.B., Assistant Surgeon to St. Bartholomew's Hospital. 8vo, with 61 Lithographic Plates and 89 Engravings. 16s.

Also.

Landmarks, Medical and Surgical. Fourth Edition. 8vo. [*In the Press.*

The Student's Guide to Surgical Anatomy. By EDWARD BELLAMY, F.R.C.S. and Member of the Board of Examiners. Third Edition. Fcap. 8vo, with 81 Engravings. 7s. 6d.

The Student's Guide to Human Osteology. By WILLIAM WARWICK WAGSTAFFE, late Assistant Surgeon to St. Thomas's Hospital. Fcap. 8vo, with 23 Plates and 66 Engravings. 10s. 6d.

The Anatomical Remembrancer ; or, Complete Pocket Anatomist. Eighth Edition. 32mo, 3s. 6d.

Diagrams of the Nerves of the Human Body,
exhibiting their Origin, Divisions, and Connections, with their Distribution to the Various Regions of the Cutaneous Surface, and to all the Muscles. By W. H. FLOWER, F.R.S., F.R.C.S. Third Edition, with 6 Plates. Royal 4to, 12s.

Atlas of Pathological Anatomy.
By Dr. LANCEREAUX. Translated by W. S. GREENFIELD, M.D., Professor of Pathology in the University of Edinburgh. Imp. 8vo, with 70 Coloured Plates, £5 5s.

A Manual of Pathological Anatomy.
By C. HANDFIELD JONES, M.B., F.R.S., and E. H. SIEVEKING, M.D., F.R.C.P. Edited by J. F. PAYNE, M.D., F.R.C.P., Lecturer on General Pathology at St. Thomas's Hospital. Second Edition. Crown 8vo, with 195 Engravings, 16s.

Post-mortem Examinations:
A Description and Explanation of the Method of Performing them, with especial reference to Medico-Legal Practice. By Prof. VIRCHOW. Translated by Dr. T. P. SMITH. Second Edition. Fcap. 8vo, with 4 Plates, 3s. 6d.

The Human Brain:
Histological and Coarse Methods of Research. A Manual for Students and Asylum Medical Officers. By W. BEVAN LEWIS, L.R.C.P. Lond., Medical Superintendent, West Riding Lunatic Asylum. 8vo, with Wood Engravings and Photographs, 8s.

Manual of Physiology:
For the use of Junior Students of Medicine. By GERALD F. YEO, M.D., F.R.C.S., Professor of Physiology in King's College, London. Crown 8vo, with 300 Engravings, 14s.

Principles of Human Physiology.
By W. B. CARPENTER, C.B., M.D., F.R.S. Ninth Edition. By HENRY POWER, M.B., F.R.C.S. 8vo, with 3 Steel Plates and 377 Wood Engravings, 31s. 6d.

Syllabus of a Course of Lectures on Physiology.
By PHILIP H. PYE-SMITH, B.A., M.D., F.R.C.P., Physician to Guy's Hospital. Crown 8vo, with Diagrams, Notes, and Tables, 5s.

A Treatise on Human Physiology.
By JOHN C. DALTON, M.D. Seventh Edition. 8vo, with 252 Engravings, 20s.

A Text-Book of Medical Physics,
for the Use of Students and Practitioners of Medicine. By JOHN C. DRAPER, M.D., LL.D., Professor of Chemistry and Physics in the Medical Department of the University of New York. With 377 Engravings. 8vo, 18s.

Histology and Histo-Chemistry of Man.
By HEINRICH FREY, Professor of Medicine in Zurich. Translated by ARTHUR E. J. BARKER, Assistant Surgeon to University College Hospital. 8vo, with 608 Engravings, 21s.

The Law of Sex.
By G. B. STARKWEATHER, F.R.G.S. With 40 Illustrative Portraits. 8vo, 16s.

Influence of Sex in Disease.
By W. ROGER WILLIAMS, F.R.C.S., Surgical Registrar to the Middlesex Hospital. 8vo, 3s. 6d.

Medical Jurisprudence:
Its Principles and Practice. By ALFRED S. TAYLOR, M.D., F.R.C.P., F.R.S. Third Edition, by THOMAS STEVENSON, M.D., F.R.C.P., Lecturer on Medical Jurisprudence at Guy's Hospital. 2 vols. 8vo, with 188 Engravings, 31s. 6d.

By the same Author.

A Manual of Medical Jurisprudence.
Tenth Edition. Crown 8vo, with 55 Engravings, 14s.

Poisons,
Also. In Relation to Medical Jurisprudence and Medicine. Third Edition. Crown 8vo, with 104 Engravings, 16s.

Lectures on Medical Jurisprudence.
By FRANCIS OGSTON, M.D., late Professor in the University of Aberdeen. Edited by FRANCIS OGSTON, Jun., M.D. 8vo, with 12 Copper Plates, 18s.

A Handy Book of Forensic Medicine and Toxicology.
By C. MEYMOTT TIDY, M.D., F.C.S., and W. BATHURST WOODMAN, M.D., F.R.C.P. 8vo, with 8 Lithographic Plates and 116 Engravings, 31s. 6d.

The Student's Guide to Medical Jurisprudence.
By JOHN ABERCROMBIE, M.D., Lecturer on Forensic Medicine to Charing Cross Hospital. Fcap. 8vo, 7s. 6d.

Microscopical Examination of Drinking Water and of Air.
By J. D. MACDONALD, M.D., F.R.S., Ex-Professor of Naval Hygiene in the Army Medical School. Second Edition. 8vo, with 25 Plates, 7s. 6d.

Dress: Its Sanitary Aspect.
A Paper read before the Brighton Social Union, Jan. 30, 1880. By BERNARD ROTH, F.R.C.S. 8vo, with 8 Plates, 2s.

Pay Hospitals and Paying Wards throughout the World.
By HENRY C. BURDETT, late Secretary to the Seamen's Hospital Society. 8vo, 7s.

By the same Author.

Cottage Hospitals — General,
Fever, and Convalescent: Their Progress, Management, and Work. Second Edition, with many Plans and Illustrations. Crown 8vo, 14s.

A Manual of Practical Hygiene.

By F. A. PARKES, M.D., F.R.S. Sixth Edition, by F. DE CHAUMONT, M.D., F.R.S., Professor of Military Hygiene in the Army Medical School. 8vo, with numerous Plates and Engravings. 18s.

A Handbook of Hygiene and Sanitary Science.

By GEO. WILSON, M.A., M.D., F.R.S.E., Medical Officer of Health for Mid-Warwickshire. Sixth Edition. Crown 8vo, with Engravings.
[*In the Press.*

By the same Author.

Healthy Life and Healthy Dwellings :

A Guide to Personal and Domestic Hygiene. Fcap. 8vo, 5s.

Sanitary Examinations

Of Water, Air, and Food. A Vade-Mecum for the Medical Officer of Health. By CORNELIUS B. FOX, M.D., F.R.C.P. Crown 8vo, with 94 Engravings, 12s. 6d.

Dangers to Health:

A Pictorial Guide to Domestic Sanitary Defects. By T. PRIDGIN TEALE, M.A., Surgeon to the Leeds General Infirmary. Fourth Edition. 8vo, with 70 Lithograph Plates (mostly coloured), 10s.

Hospitals, Infirmaries, and Dispensaries:

Their Construction, Interior Arrangement, and Management; with Descriptions of existing Institutions, and 74 Illustrations. By F. OPPERT, M.D., M.R.C.P.L. Second Edition. Royal 8vo, 12s.

Hospital Construction and Management.

By F. J. MOUAT, M.D., Local Government Board Inspector, and H. SAXON SNELL, Fell. Roy. Inst. Brit. Architects. In 2 Parts, 4to, 15s. each; or, the whole work bound in half calf, with large Map, 54 Lithographic Plates, and 27 Woodcuts, 35s.

Manual of Anthropometry:

A Guide to the Measurement of the Human Body, containing an Anthropometrical Chart and Register, a Systematic Table of Measurements, &c. By CHARLES ROBERTS, F.R.C.S. 8vo, with numerous Illustrations and Tables, 8s. 6d.

By the same Author.

Detection of Colour-Blindness and Imperfect Eyesight.

8vo, with a Table of Coloured Wools, and Sheet of Test-types, 5s.

Illustrations of the Influence of the Mind upon the Body in Health and Disease :

Designed to elucidate the Action of the Imagination. By DANIEL HACK TUKE, M.D., F.R.C.P., LL.D. Second Edition. 2 vols. crown 8vo, 15s.

By the same Author.

Sleep-Walking and Hypnotism.

8vo, 5s.

A Manual of Psychological Medicine.

With an Appendix of Cases. By JOHN C. BUCKNILL, M.D., F.R.S., and D. HACK TUKE, M.D., F.R.C.P. Fourth Edition. 8vo, with 12 Plates (30 Figures) and Engravings, 25s.

Mental Diseases.

Clinical Lectures. By T. S. CLOUSTON, M.D., F.R.C.P. Edin., Lecturer on Mental Diseases in the University of Edinburgh. With 8 Plates (6 Coloured). Crown 8vo, 12s. 6d.

The Student's Guide to the Practice of Midwifery.

By D. LLOYD ROBERTS, M.D., F.R.C.P., Lecturer on Clinical Midwifery and Diseases of Women at Owen's College; Obstetric Physician to the Manchester Royal Infirmary. Third Edition. Fcap. 8vo, with 2 Coloured Plates and 127 Wood Engravings, 7s. 6d.

Handbook of Midwifery for Midwives:

By J. E. BURTON, L.R.C.P. Lond., Surgeon to the Hospital for Women, Liverpool. Second Edition. With Engravings. Fcap. 8vo, 6s.

Lectures on Obstetric Operations:

Including the Treatment of Hæmorrhage, and forming a Guide to the Management of Difficult Labour. By ROBERT BARNES, M.D., F.R.C.P., Consulting Obstetric Physician to St. George's Hospital. Fourth Edition. 8vo, with 121 Engravings, 12s. 6d.

By the same Author.

A Clinical History of Medical and Surgical Diseases of Women.

Third Edition. 8vo, with Engravings.
[*In the Press.*

Clinical Lectures on Diseases of Women:

Delivered in St. Bartholomew's Hospital, by J. MATTHEWS DUNCAN, M.D., F.R.C.P., F.R.S.E. Third Edition. 8vo.
[*In the Press.*

By the same Author.

Sterility in Woman.

Being the Gulstonian Lectures, delivered in the Royal College of Physicians, in Feb., 1883. 8vo, 6s.

Notes on Diseases of Women:

Specially designed to assist the Student in preparing for Examination. By J. J. REYNOLDS, L.R.C.P., M.R.C.S. Second Edition. Fcap. 8vo, 2s. 6d.

By the same Author.

Notes on Midwifery:

Specially designed for Students preparing for Examination. Fcap. 8vo, 4s.

Schroeder's Manual of Midwifery,

including the Pathology of Pregnancy and the Puerperal State. Translated by CHARLES H. CARTER, B.A., M.D. 8vo, with Engravings, 12s. 6d.

The Student's Guide to the Diseases of Women.
By ALFRED L. GALABIN, M.D., F.R.C.P., Obstetric Physician to Guy's Hospital. Third Edition. Fcap. 8vo, with 78 Engravings, 7s. 6d.

Dysmenorrhœa, its Pathology and Treatment.
By HEYWOOD SMITH, M.D. Oxon., late Physician to the Hospital for Women, &c. Crown 8vo, with Engravings, 4s. 6d.

Obstetric Aphorisms:
For the Use of Students commencing Midwifery Practice. By JOSEPH G. SWAYNE, M.D. Eighth Edition. Fcap. 8vo, with Engravings, 3s. 6d.

Influence of Posture on Women
in Gynecic and Obstetric Practice. By J. H. AVELING, M.D., Physician to the Chelsea Hospital for Women. 8vo, 6s.

By the same Author.

The Chamberlens and the Midwifery Forceps:
Memorials of the Family, and an Essay on the Invention of the Instrument. 8vo, with Engravings, 7s. 6d.

A Handbook of Uterine Therapeutics,
and of Diseases of Women. By E. J. TILT, M.D., M.R.C.P. Fourth Edition. Post 8vo, 10s.

By the same Author.

The Change of Life
In Health and Disease: A Clinical Treatise on the Diseases of the Nervous System incidental to Women at the Decline of Life. Fourth Edition. 8vo, 10s. 6d.

The Principles and Practice of Gynæcology.
By THOMAS ADDIS EMMET, M.D., Surgeon to the Woman's Hospital, New York. Third Edition. Royal 8vo, with 150 Engravings, 24s.

Diseases of the Uterus, Ovaries, and Fallopian Tubes:
A Practical Treatise by A. COURTY, Professor of Clinical Surgery, Montpellier. Translated from Third Edition by his Pupil, AGNES McLAREN, M.D., M.K.Q.C.P.I., with Preface by J. MATTHEWS DUNCAN, M.D., F.R.C.P. 8vo, with 424 Engravings, 24s.

The Female Pelvic Organs:
Their Surgery, Surgical Pathology, and Surgical Anatomy. In a Series of Coloured Plates taken from Nature; with Commentaries, Notes, and Cases. By HENRY SAVAGE, M.D., F.R.C.S., Consulting Officer of the Samaritan Free Hospital. Fifth Edition. Roy. 4to, with 17 Lithographic Plates (15 coloured) and 52 Woodcuts, £1 15s.

Backward Displacements of the Uterus and Prolapsus Uteri:
Treatment by the New Method of Shortening the Round Ligaments. By WILLIAM ALEXANDER, M.D., M.Ch.Q.U.I., F.R.C.S., Surgeon to the Liverpool Infirmary. Crown 8vo, with Engravings, 3s. 6d.

Ovarian and Uterine Tumours:
Their Pathology and Surgical Treatment. By Sir T. SPENCER WELLS, Bart., F.R.C.S., Consulting Surgeon to the Samaritan Hospital. 8vo, with Engravings, 21s.

By the same Author.

Abdominal Tumours:
Their Diagnosis and Surgical Treatment. 8vo, with Engravings, 3s. 6d.

West on the Diseases of Women.
Fourth Edition, revised by the Author, with numerous Additions by J. MATTHEWS DUNCAN, M.D., F.R.C.P., F.R.S.E., Obstetric Physician to St. Bartholomew's Hospital. 8vo, 16s.

The Student's Guide to Diseases of Children.
By JAS. F. GOODHART, M.D., F.R.C.P., Assistant Physician to Guy's Hospital; Physician to the Evelina Hospital for Sick Children. Fcap. 8vo, 10s. 6d.

The Wasting Diseases of Infants and Children.
By EUSTACE SMITH, M.D., Physician to the King of the Belgians, Physician to the East London Hospital for Children. Fourth Edition. Post 8vo, 8s. 6d.

By the same Author.

Diseases in Children:
A Practical Treatise. 8vo, 22s.

Also.

Clinical Studies of Disease in Children.
Second Edition. Post 8vo. [*In the Press.*

A Practical Manual of the Diseases of Children.
With a Formulary. By EDWARD ELLIS, M.D. Fourth Edition. Crown 8vo, 10s.

By the same Author.

A Manual of what every Mother
should know. Fcap. 8vo, 1s. 6d.

A Manual for Hospital Nurses
and others engaged in Attending on the Sick. By EDWARD J. DOMVILLE, Surgeon to the Exeter Lying-in Charity. Fifth Edition. Crown 8vo, 2s. 6d.

A Manual of Nursing, Medical and Surgical.
By CHARLES J. CULLINGWORTH, M.D., Physician to St. Mary's Hospital, Manchester. Second Edition. Fcap. 8vo, with Engravings, 3s. 6d.

By the same Author.

A Short Manual for Monthly Nurses.
Fcap. 8vo, 1s. 6d.

Notes on Fever Nursing.
By J. W. ALLAN, M.B., Physician, Superintendent Glasgow Fever Hospital. Crown 8vo, with Engravings, 2s. 6d.

By the same Author.

Outlines of Infectious Diseases:
For the use of Clinical Students. Fcap. 8vo.

Hospital Sisters and their Duties.
By EVA C. E. LÜCKES, Matron to the London Hospital. Crown 8vo, 2s. 6d.

Diseases of Children.

For Practitioners and Students. By W. H. DAY, M.D., Physician to the Samaritan Hospital. Second Edition. Crown 8vo, 12s. 6d.

Infant Feeding and its Influence on Life ;

By C. II. F. ROUTH, M.D., Physician to the Samaritan Hospital. Third Edition. Fcap. 8vo, 7s. 6d.

Manual of Botany:

Including the Structure, Functions, Classification, Properties, and Uses of Plants. By ROBERT BENTLEY, Professor of Botany in King's College and to the Pharmaceutical Society. Fourth Edition. Crown 8vo, with 1,185 Engravings, 15s.

By the same Author.

The Student's Guide to Structural, Morphological, and Physiological Botany.

With 660 Engravings. Fcap. 8vo, 7s. 6d.

Also.

The Student's Guide to Systematic Botany,

including the Classification of Plants and Descriptive Botany. Fcap. 8vo, with 350 Engravings, 3s. 6d.

Medicinal Plants :

Being descriptions, with original figures, of the Principal Plants employed in Medicine, and an account of their Properties and Uses. By Prof. BENTLEY and Dr. H. TRIMEN. In 4 vols., large 8vo, with 306 Coloured Plates, bound in Half Morocco, Gilt Edges, £11 11s.

The National Dispensatory :

Containing the Natural History, Chemistry, Pharmacy, Actions and Uses of Medicines. By ALFRED STILLÉ, M.D., LL.D., and JOHN M. MAISCH, Ph.D. Third Edition. 8vo, with 311 Engravings, 36s.

Royle's Manual of Materia Medica and Therapeutics.

Sixth Edition. By JOHN HARLEY, M.D., Physician to St. Thomas's Hospital. Crown 8vo, with 139 Engravings, 15s.

Materia Medica and Therapeutics.

By CHARLES D. F. PHILLIPS, M.D., F.R.S. Edin., late Lecturer on Materia Medica and Therapeutics at the Westminster Hospital Medical School.

Vol. 1—Vegetable Kingdom. 8vo, 15s.
Vol. 2—Inorganic Substances. 8vo, 21s.

Binz's Elements of Therapeutics :

A Clinical Guide to the Action of Drugs. Translated by E. I. SPARKS, M.B., F.R.C.P. Crown 8vo, 8s. 6d.

Materia Medica.

A Manual for the use of Students. By ISAMBARD OWEN, M.D., F.R.C.P., Lecturer on Materia Medica, &c., to St. George's Hospital. Crown 8vo, 6s.

The Student's Guide to Materia Medica and Therapeutics.

By JOHN C. THOROWGOOD, M.D., F.R.C.P. Second Edition. Fcap. 8vo, 7s.

The Pharmacopœia of the London Hospital.

Compiled under the direction of a Committee appointed by the Hospital Medical Council. Fcap. 8vo, 3s.

A Companion to the British Pharmacopœia.

By PETER SQUIRE, F.L.S., assisted by his Sons, P. W. and A. II. SQUIRE. 13th Edition. 8vo, 10s. 6d.

By the same Authors.

The Pharmacopœias of the London Hospitals,

arranged in Groups for Easy Reference and Comparison. Fifth Edition. 18mo, 6s.

The Prescriber's Pharmacopœia:

The Medicines arranged in classes according to their Action, with their Composition and Doses. Sixth Edition. By NESTOR J. C. TIRARD, M.D., M.R.C.P., Professor of Materia Medica and Therapeutics in King's College, London. 32mo, bound in leather, 3s.

Bazaar Medicines of India,

And Common Medical Plants : With Full Index of Diseases, indicating their Treatment by these and other Agents procurable throughout India, &c. By E. J. WARING, C.I.E., M.D., F.R.C.P. Fourth Edition. Fcap. 8vo, 5s.

Tropical Dysentery and Chronic Diarrhœa

— Liver Abscess — Malarial Cachexia—Insolation—with other forms of Tropical Diseases, &c. By Sir JOSEPH FAYRER, K.C.S.I., M.D. 8vo, 15s.

By the same Author.

Climate and Fevers of India,

with a series of Cases (Croonian Lectures, 1882). 8vo, with 17 Temperature Charts, 12s.

Family Medicine for India.

A Manual. By WILLIAM J. MOORE, M.D., C.I.E., Honorary Surgeon to the Viceroy of India. Published under the Authority of the Government of India. Fourth Edition. Post 8vo, with 64 Engravings, 12s.

By the same Author.

Health-Resorts for Tropical Invalids,

in India, at Home, and Abroad. Post 8vo, 5s.

Spirillum Fever

(Synonyms, Famine or Relapsing Fever), as seen in Western India. By H. VANDYKE CARTER, M.D., F.R.C.P., Surgeon-Major I.M.D. 8vo, with Plates, 21s.

Clinical Medicine :

A Systematic Treatise on the Diagnosis and Treatment of Disease. By AUSTIN FLINT, M.D., Professor of Medicine in the Bellevue Hospital Medical College. 8vo, 20s.

By the same Author.

Phthisis :

In a series of Clinical Studies. 8vo, 16s.

The Principles and Practice of Medicine.
By C. HILTON FAGGE, M.D. Edited by P. H. PYE-SMITH, M.D., F.R.C.P., Physician to, and Lecturer on Medicine at, Guy's Hospital. 2 vols. 8vo, 1860 pp. Cloth, 36s. ; Half Persian, 42s.

The Student's Guide to the Practice of Medicine.
By MATTHEW CHARTERIS, M.D., Professor of Materia Medica in the University of Glasgow. Fourth Edition. Fcap. 8vo, with Engravings on Copper and Wood. 9s.

Hooper's Physicians' Vade-Mecum.
A Manual of the Principles and Practice of Physic. Tenth Edition. By W. A. GUY, F.R.C.P., F.R.S., and J. HARLEY, M.D., F.R.C.P. With 118 Engravings. Fcap. 8vo, 12s. 6d.

The Student's Guide to Clinical Medicine and Case-Taking.
By FRANCIS WARNER, M.D., F.R.C.P., Assistant Physician to the London Hospital. Second edition. Fcap. 8vo, 5s.

The Student's Guide to Medical Diagnosis.
By SAMUEL FENWICK, M.D., F.R.C.P., Physician to the London Hospital. Sixth Edition. Fcap. 8vo, with 114 Engravings, 7s.

By the same Author.

The Student's Outlines of Medical Treatment.
Second Edition. Fcap. 8vo, 7s.

Also.

On Chronic Atrophy of the Stomach,
and on the Nervous Affections of the Digestive Organs. 8vo, 8s.

How to Examine the Chest :
Being a Practical Guide for the use of Students. By SAMUEL WEST, M.D., F.R.C.P., Physician to the City of London Hospital for Diseases of the Chest ; Medical Tutor and Registrar at St. Bartholomew's Hospital. With 42 Engravings. Fcap. 8vo, 5s.

The Contagiousness of Pulmonary Consumption, and its Antiseptic Treatment.
By J. BURNEY YEO, M.D., Physician to King's College Hospital. Crown 8vo, 3s. 6d.

The Operative Treatment of Intra-thoracic Effusion.
Fothergillian Prize Essay. By NORMAN PORRITT, L.R.C.P. Lond., M.R.C.S. With Engravings. Crown 8vo, 6s.

The Spectroscope in Medicine.
By CHARLES A. MACMUNN, B.A., M.D. 8vo, with 3 Chromo-lithographic Plates of Physiological and Pathological Spectra, and 13 Engravings, 9s.

Notes on Asthma :
Its Forms and Treatment. By JOHN C. THOROWGOOD, M.D., Physician to the Hospital for Diseases of the Chest. Third Edition. Crown 8vo, 4s. 6d.

The Microscope in Medicine.
By LIONEL S. BEALE, M.B., F.R.S., Physician to King's College Hospital. Fourth Edition. 8vo, with 86 Plates, 21s.

Also.

On Slight Ailments :
Their Nature and Treatment. Second Edition. 8vo, 5s.

Diseases of the Chest :
Contributions to their Clinical History, Pathology, and Treatment. By A. T. HOUGHTON WATERS, M.D., Physician to the Liverpool Royal Infirmary. Second Edition. 8vo, with Plates, 15s.

Winter Cough
(Catarrh, Bronchitis, Emphysema, Asthma). By HORACE DOBELL, M.D., Consulting Physician to the Royal Hospital for Diseases of the Chest. Third Edition. 8vo, with Coloured Plates, 10s. 6d.

By the same Author.

Loss of Weight, Blood-Spitting, and Lung Disease.
Second Edition, to which is added Part VI., "On the Functions and Diseases of the Liver." 8vo, with Chromo-lithograph, 10s. 6d.

Also.

The Mont Dore Cure, and the Proper Way to Use it.
8vo, 7s. 6d.

Croonian Lectures on Some Points in the Pathology and Treatment of Typhoid Fever.
By WILLIAM CAYLEY, M.D., F.R.C.P., Physician to the Middlesex and the London Fever Hospitals. Crown 8vo, 4s. 6d.

Diseases of the Heart and Aorta :
Clinical Lectures. By G. W. BALFOUR, M.D., F.R.C.P., F.R.S. Edin., late Senior Physician and Lecturer on Clinical Medicine, Royal Infirmary, Edinburgh. Second Edition. 8vo, with Chromo-lithograph and Wood Engravings, 12s. 6d.

Manual of the Physical Diagnosis of Diseases of the Heart,
including the use of the Sphygmograph and Cardiograph. By A. E. SANSOM, M.D., F.R.C.P., Assistant Physician to the London Hospital. Third Edition. Fcap. 8vo, with 48 Engravings, 7s. 6d.

Visceral Neuroses :
Being the Gulstonian Lectures on Neuralgia of the Stomach, and Allied Disorders. By T. CLIFFORD ALLBUTT, M.A., M.D. Cantab., F.R.S., F.R.C.P., Consulting Physician to the Leeds General Infirmary. 8vo, 4s. 6d.

Nervous Diseases :
Their Description and Treatment. A Manual for Students and Practitioners of Medicine. By ALLEN M. HAMILTON, M.D., Physician at the Epileptic and Paralytic Hospital, New York. Second Edition. Royal 8vo, with 72 Engravings, 16s.

Medical Ophthalmoscopy :
A Manual and Atlas. By WILLIAM R. GOWERS, M.D., F.R.C.P., Assistant Professor of Clinical Medicine in University College, and Senior Assistant Physician to the Hospital. Second Edition, with Coloured Autotype and Lithographic Plates and Woodcuts. 8vo, 18s.

By the same Author.

Epilepsy, and other Chronic
Convulsive Diseases : Their Causes, Symptoms, and Treatment. 8vo, 10s. 6d.

Also.

Pseudo-Hypertrophic Muscular
Paralysis : A Clinical Lecture. 8vo, with Engravings and Plate, 3s. 6d.

Also.

Diagnosis of Diseases of the
Spinal Cord. Third Edition. 8vo, with Engravings, 4s. d.

Also.

Diagnosis of Diseases of the
Brain. 8vo, with Engravings, 7s. 6d.

Diseases of the Nervous System.
Lectures delivered at Guy's Hospital. By SAMUEL WILKS, M.D., F.R.S. Second Edition. 8vo, 18s.

Diseases of the Nervous System:
Especially in Women. By S. WEIR MITCHELL, M.D., Physician to the Philadelphia Infirmary for Diseases of the Nervous System. Second Edition. 8vo, with 5 plates, 8s.

Nerve Vibration and Excitation,
as Agents in the Treatment of Functional Disorder and Organic Disease. By J. MORTIMER GRANVILLE, M.D. 8vo, 5s.

By the same Author.

Gout in its Clinical Aspects.
Crown 8vo, 6s.

Regimen to be adopted in Cases
of Gout. By WILHELM EBSTEIN, M.D., Professor of Clinical Medicine in Göttingen. Translated by JOHN SCOTT, M.A., M.B. 8vo, 2s. 6d.

Notes on Rheumatism.
By JULIUS POLLOCK, M.D., F.R.C.P., Senior Physician to the Charing Cross Hospital. Second Edition. Fcap. 8vo, with Engravings, 3s. 6d.

Diseases of the Liver :
With and without Jaundice. By GEORGE HARLEY, M.D., F.R.C.P., F.R.S. 8vo, with 2 Plates and 36 Engravings, 21s.

Gout, Rheumatism,
And the Allied Affections ; with Chapters on Longevity and Sleep. By PETER HOOD, M.D. Third Edition. Crown 8vo, 7s. 6d.

Diseases of the Nervous System.
Clinical Lectures. By THOMAS BUZZARD, M.D., F.R.C.P., Physician to the National Hospital for the Paralysed and Epileptic. With Engravings, 8vo. 15s.

Diseases of the Stomach :
The Varieties of Dyspepsia, their Diagnosis and Treatment. By S. O. HABERSHON, M.D., F.R.C.P. Third Edition. Crown 8vo, 5s.

By the same Author.

Pathology of the Pneumo-
gastric Nerve : Being the Lumleian Lectures for 1876. Second edition. Post 8vo, 4s.

Also.

Diseases of the Abdomen,
Comprising those of the Stomach and other parts of the Alimentary Canal, Œsophagus, Cæcum, Intestines, and Peritoneum. Third Edition. 8vo, with 5 Plates, 21s.

Also.

Diseases of the Liver,
Their Pathology and Treatment. Lettsomian Lectures. Second Edition. Post 8vo, 4s.

Acute Intestinal Strangulation,
And Chronic Intestinal Obstruction (Mode of Death from). By THOMAS BRYANT, F.R.C.S., Senior Surgeon to Guy's Hospital. 8vo, 3s.

A Treatise on the Diseases of
the Nervous System. By JAMES ROSS, M.D., F.R.C.P., Assistant Physician to the Manchester Royal Infirmary. Second Edition. 2 vols. 8vo, with Lithographs, Photographs, and 332 Woodcuts, 52s. 6d.

By the same Author.

Handbook of the Diseases of
the Nervous System. 8vo, with 184 Engravings, 18s.

Food and Dietetics,
Physiologically and Therapeutically Considered. By F. W. PAVY, M.D., F.R.S., Physician to Guy's Hospital. Second Edition. 8vo, 15s.

By the same Author.

Croonian Lectures on Certain
Points connected with Diabetes. 8vo, 4s. 6d.

Headaches :
Their Nature, Causes, and Treatment. By W. H. DAY, M.D., Physician to the Samaritan Hospital. Third Edition. Crown 8vo, with Engravings, 6s. 6d.

On Megrim, Sick Headache, and
some Allied Disorders : A Contribution to the Pathology of Nerve Storms. By E. LIVEING, M.D., F.R.C.P. 8vo, 15s.

The Principal Southern and
Swiss Health-Resorts : their Climate and Medical Aspect. By WILLIAM MARCET, M.D., F.R.C.P., F.R.S. With Illustrations. Crown 8vo, 7s. 6d.

Principal Health-Resorts
Of Europe and Africa, and their Use in the Treatment of Chronic Diseases. By T. M. MADDEN, M.D. 8vo, 10s.

Health Resorts at Home and Abroad.
By MATTHEW CHARTERIS, M.D., Physician to the Glasgow Royal Infirmary. Crown 8vo, with Map, 4s. 6d.

Winter and Spring
On the Shores of the Mediterranean. By HENRY BENNET, M.D. Fifth Edition. Post 8vo, with numerous Plates, Maps, and Engravings, 12s. 6d.

By the same Author.

Treatment of Pulmonary Consumption
by Hygiene, Climate, and Medicine. Third Edition. 8vo, 7s. 6d.

Also.

Nutrition in Health and Disease.
Third (Library) Edition, 8vo, 5s. ; Cheap Edition, fcap. 8vo, 2s. 6d.

The Riviera :
Sketches of the Health-Resorts of the Coast of France and Italy, from Hyères to Spezia : its Medical Aspect and Value, &c. By EDWARD I. SPARKS, M.B., F.R.C.P. Crown 8vo, 8s. 6d.

Medical Guide to the Mineral Waters of France and its Wintering Stations.
With a Special Map. By A. VINTRAS, M.D., Physician to the French Embassy, and to the French Hospital, London. Crown 8vo, 8s.

The Ocean as a Health-Resort :
A Practical Handbook of the Sea, for the use of Tourists and Health-Seekers. By WILLIAM S. WILSON, L.R.C.P. Second Edition, with Chart of Ocean Routes, &c. Crown 8vo, 7s. 6d.

Electricity and its Manner of Working in the Treatment of Disease.
By WM. E. STEAVENSON, M.D., Physician and Electrician to St Bartholomew's Hospital. 8vo, 4s. 6d.

Mechanical Exercise a Means of Cure :
Being a Description of the Zander Institute, London ; its History, Appliances, Scope, and Object. Edited by the Medical Officer of the Institution. Crown 8vo, with 24 Engravings, 2s. 6d.

On Dislocations and Fractures.
By JOSEPH MACLISE, F.R.C.S. Uniform with "Surgical Anatomy." 36 folio Plates and Text. Cloth, £2 10s.

Ambulance Handbook for Volunteers and Others.
By J. ARDAVON RAYE, L.K. & Q.C.P.I., L.R.C.S.I., late Surgeon to H.B.M. Transport No. 14, Zulu Campaign, and Surgeon E.I.R. Rifles. 8vo, with 16 Plates (50 figures), 3s. 6d.

Surgical Emergencies :
Together with the Emergencies Attendant on Parturition and the Treatment of Poisoning. By PAUL SWAIN, F.R.C.S., Surgeon to the South Devon and East Cornwall Hospital. Third Edition. Crown 8vo, with 117 Engravings, 5s.

Handbook of Medical and Surgical Electricity.
By HERBERT TIBBITS, M.D., F.R.C.P.E., Senior Physician to the West London Hospital for Paralysis and Epilepsy. Second Edition. 8vo, with 95 Engravings, 9s.

By the same Author.

How to Use a Galvanic Battery in Medicine and Surgery.
Third Edition. 8vo, with Engravings, 4s.

Also.

A Map of Ziemssen's Motor Points of the Human Body :
A Guide to Localised Electrisation. Mounted on Rollers, 35 × 21. With 20 Illustrations, 5s.

Operative Surgery in the Calcutta Medical College Hospital.
Statistics, Cases, and Comments. By KENNETH McLEOD, A.M., M.D., F.R.C.S.E., Surgeon-Major, Indian Medical Service, Professor of Surgery in Calcutta Medical College. 8vo, with Illustrations, 12s. 6d.

A Course of Operative Surgery.
By CHRISTOPHER HEATH, Surgeon to University College Hospital. Second Edition. With 20 coloured Plates (180 figures) from Nature, by M. LÉVEILLÉ, and several Woodcuts. Large 8vo, 30s.

By the same Author.

The Student's Guide to Surgical Diagnosis.
Second Edition. Fcap. 8vo, 6s. 6d.

Also.

Manual of Minor Surgery and Bandaging.
For the use of House-Surgeons, Dressers, and Junior Practitioners. Seventh Edition. Fcap. 8vo, with 129 Engravings, 6s.

Also.

Injuries and Diseases of the Jaws.
Third Edition. 8vo, with Plate and 206 Wood Engravings, 14s.

Injuries and Diseases of the Neck and Head, the Genito-Urinary Organs, and the Rectum.
Hunterian Lectures, 1885. By EDWARD LUND, F.R.C.S., Professor of Surgery in the Owen's College, Manchester. 8vo, with Plates and Engravings, 4s. 6d.

Surgical Enquiries :
Including the Hastings Essay on Shock, the Treatment of Inflammations, and numerous Clinical Lectures. By FURNEAUX JORDAN, F.R.C.S., Professor of Surgery, Queen's College, Birmingham. Second Edition, with numerous Plates. Royal 8vo, 12s. 6d.

Outlines of Surgery and Surgical Pathology.
By F. LE GROS CLARK, F.R.S., assisted by W. W. WAGSTAFFE, F.R.C.S. Second Edition. 8vo, 10s. 6d.

The Practice of Surgery:

A Manual. By THOMAS BRYANT, Surgeon to Guy's Hospital. Fourth Edition. 2 vols. crown 8vo, with 750 Engravings (many being coloured), and including 6 chromo-lithographic plates, 32s.

The Surgeon's Vade-Mecum:

A Manual of Modern Surgery. By ROBERT DRUITT, F.R.C.S. Twelfth Edition. Fcap. 8vo, with Engravings. [In the Press.

Regional Surgery:

Including Surgical Diagnosis. A Manual for the use of Students. By F. A. SOUTHAM, M.A., M.B., F.R.C.S., Assistant Surgeon to the Manchester Royal Infirmary. Part I. The Head and Neck. Crown 8vo, 6s. 6d. — Part II. The Upper Extremity and Thorax. Crown 8vo, 7s. 6d.

Illustrations of Clinical Surgery.

By JONATHAN HUTCHINSON, F.R.S., Senior Surgeon to the London Hospital. In occasional fasciculi. I. to XVIII., 6s. 6d. each. Fasciculi I. to X. bound, with Appendix and Index, £3 10s.

By the same Author.

Pedigree of Disease:

Being Six Lectures on Temperament, Idiosyncrasy, and Diathesis. 8vo, 5s.

Treatment of Wounds and Fractures.

Clinical Lectures. By SAMPSON GAMGEE. F.R.S.E., Surgeon to the Queen's Hospital, Birmingham. Second Edition. 8vo, with 40 Engravings, 10s.

Injuries of the Spine and Spinal Cord, and NERVOUS SHOCK,

in their Surgical and Medico-Legal Aspects. By HERBERT W. PAGE, M.C. Cantab., F.R.C.S., Surgeon to St. Mary's Hospital. Second Edition, post 8vo, 10s.

Lectures on Orthopædic Surgery.

By BERNARD E. BRODHURST, F.R.C.S., Surgeon to the Royal Orthopædic Hospital. Second Edition. 8vo, with Engravings, 12s. 6d.

By the same Author.

On Anchylosis, and the Treatment for the Removal of Deformity and the Restoration of Mobility in Various Joints.

Fourth Edition. 8vo, with Engravings, 5s.

Also.

Curvatures and Diseases of the Spine.

Third Edition. 8vo, with Engravings, 6s.

Orthopædic Surgery,

And Diseases of the Joints. By L. A. SAYRE, M.D., Professor of Orthopædic Surgery in Bellevue Hospital Medical College. Second Edition. 8vo, with Coloured Plate and 324 Engravings, 21s.

Face and Foot Deformities.

By FREDERICK CHURCHILL, C.M., Surgeon to the Victoria Hospital for Children. 8vo, with Plates and Illustrations, 10s. 6d.

Fractures:

A Treatise. By LEWIS A. STIMSON, B.A., M.D., Professor of Surgical Pathology in the University of New York. 8vo, with 360 Engravings, 21s.

Clubfoot:

Its Causes, Pathology, and Treatment. By WM. ADAMS, F.R.C.S., Surgeon to the Great Northern Hospital. Second Edition. 8vo, with 106 Engravings and 6 Lithographic Plates, 15s.

By the same Author.

On Contraction of the Fingers,

and its Treatment by Subcutaneous Operation; and on Obliteration of Depressed Cicatrices, by the same Method. 8vo, with 30 Engravings, 4s. 6d.

Also.

Lateral and other Forms of Curvature of the Spine:

Their Pathology and Treatment. Second Edition. 8vo, with 5 Lithographic Plates and 72 Wood Engravings, 10s. 6d.

Spinal Curvatures:

Treatment by Extension and Jacket; with Remarks on some Affections of the Hip, Knee, and Ankle-joints. By H. MACNAUGHTON JONES, M.D., F.R.C.S.I. and Edin. Post 8vo, with 63 Engravings, 4s. 6d.

On Diseases and Injuries of the Eye:

A Course of Systematic and Clinical Lectures to Students and Medical Practitioners. By J. R. WOLFE, M.D., F.R.C.S.E., Lecturer on Ophthalmic Medicine and Surgery in Anderson's College, Glasgow. With 10 Coloured Plates and 157 Wood Engravings. 8vo, £1 1s.

The General Practitioner's Guide to the Diseases and Injuries of the Eye and Eyelids.

By LOUIS H. TOSSWILL, B.A., M.B. Cantab., M.R.C.S., Surgeon to the Exeter Eye Infirmary. Fcap. 8vo, 2s. 6d.

Hints on Ophthalmic Out-Patient Practice.

By CHARLES HIGGENS, Ophthalmic Surgeon to Guy's Hospital. Third Edition. Fcap. 8vo, 3s. [In the Press.

The Electro-Magnet,

And its Employment in Ophthalmic Surgery. By SIMEON SNELL, Ophthalmic Surgeon to the Sheffield General Infirmary, &c. Crown 8vo, 3s. 6d.

Manual of the Diseases of the Eye.

By CHARLES MACNAMARA, F.R.C.S., Surgeon to Westminster Hospital. Fourth Edition. Crown 8vo, with 4 Coloured Plates and 66 Engravings, 10s. 6d.

The Student's Guide to Diseases
of the Eye. By EDWARD NETTLESHIP, F.R.C.S., Ophthalmic Surgeon to St. Thomas's Hospital. Third Edition. Fcap. 8vo, with 150 Engravings and a Set of Coloured Papers illustrating Colour-Blindness, 7s. 6d.

A Manual of the Principles and
Practice of Ophthalmic Medicine and Surgery. By T. WHARTON JONES, F.R.C.S., F.R.S. Third Edition. Fcap. 8vo, with 9 Coloured Plates and 173 Engravings, 12s. 6d.

Atlas of Ophthalmoscopy.
Composed of 12 Chromo-lithographic Plates (59 Figures drawn from nature) and Explanatory Text. By RICHARD LIEBREICH, M.R.C.S. Translated by H. ROSBOROUGH SWANZY, M.B. Third edition, 4to, 40s.

Glaucoma :
Its Causes, Symptoms, Pathology, and Treatment. By PRIESTLEY SMITH, M.R.C.S., Ophthalmic Surgeon to the Queen's Hospital, Birmingham. 8vo, with Lithographic Plates, 10s. 6d.

Refraction of the Eye :
A Manual for Students. By GUSTAVUS HARTRIDGE, F.R.C.S., Assistant Physician to the Royal Westminster Ophthalmic Hospital. Second Edition. Crown 8vo, with Lithographic Plate and 94 Woodcuts, 5s. 6d.

Hare-Lip and Cleft Palate.
By FRANCIS MASON, F.R.C.S., Surgeon to St. Thomas's Hospital. 8vo, with 66 Engravings, 6s.

By the same Author.

The Surgery of the Face.
8vo, with 100 Engravings, 7s. 6d.

A Practical Treatise on Aural
Surgery. By H. MACNAUGHTON JONES, M.D., Professor of the Queen's University in Ireland, late Surgeon to the Cork Ophthalmic and Aural Hospital. Second Edition. Crown 8vo, with 63 Engravings, 8s. 6d.

By the same Author.

Atlas of Diseases of the Mem-
brana Tympani. In Coloured Plates, containing 62 Figures, with Text. Crown 4to, 21s.

Diseases and Injuries of the
Ear. By Sir W. B. DALBY, F.R.C.S., Aural Surgeon to St. George's Hospital. Third Edition. Crown 8vo, with Engravings, 7s. 6d.

Lectures on Syphilis of the
Larynx (Lesions of the Secondary and Intermediate Stages). By W. M. WHISTLER, M.D., Physician to the Hospital for Diseases of the Throat. Post 8vo, 4s.

Diseases of the Throat and
Nose : A Manual. By MORELL MACKENZIE, M.D. Lond., Senior Physician to the Hospital for Diseases of the Throat. Vol. I. Diseases of the Pharynx, Larynx, and Trachea. Post 8vo, with 112 Engravings, 12s. 6d.
Vol. II. Diseases of the Nose and Naso-Pharynx ; with a Section on Diseases of the Œsophagus. Post 8vo, with 93 Engravings, 12s. 6d.

By the same Author.

Diphtheria :
Its Nature and Treatment, Varieties, and Local Expressions. 8vo, 5s.

Endemic Goitre or Thyreocele :
Its Etiology, Clinical Characters, Pathology, Distribution, Relations to Cretinism, Myxœdema, &c., and Treatment. By WILLIAM ROBINSON, M.D. 8vo, 5s.

Sore Throat :
Its Nature, Varieties, and Treatment. By PROSSER JAMES, M.D., Physician to the Hospital for Diseases of the Throat. Fifth Edition. Post 8vo, with Coloured Plates and Engravings, 6s. 6d.

The Ear :
Its Anatomy, Physiology, and Diseases. By C. H. BURNETT, A.M., M.D., Aural Surgeon to the Presbyterian Hospital, Philadelphia. Second Edition, 8vo, with 107 Engravings, 18s.

A Treatise on Vocal Physio-
logy and Hygiene, with especial reference to the Cultivation and Preservation of the Voice. By GORDON HOLMES, M.D., Physician to the Municipal Throat and Ear Infirmary. Second Edition, with Engravings. Crown 8vo, 6s. 6d.

By the same Author.

A Guide to the Use of the
Laryngoscope in General Practice. Crown 8vo, with Engravings, 2s. 6d.

A System of Dental Surgery.
By JOHN TOMES, F.R.S., and C. S. TOMES, M.A., F.R.S. Third Edition. Fcap. 8vo, with many Engravings.
[In the Press.

Dental Anatomy, Human and
Comparative : A Manual. By CHARLES S. TOMES, M.A., F.R.S. Second Edition. Crown 8vo, with 191 Engravings, 12s. 6d.

The Student's Guide to Dental
Anatomy and Surgery. By HENRY SEWILL, M.R.C.S., L.D.S. Second Edition. Fcap. 8vo, with 78 Engravings, 5s. 6d.

Mechanical Dentistry in Gold
and Vulcanite. By F. H. BALKWILL, L.D.S.R.C.S. 8vo, with 2 Lithographic Plates and 57 Engravings, 10s.

Principles and Practice of Dentistry :
including Anatomy, Physiology, Pathology, Therapeutics, Dental Surgery, and Mechanism. By CHAPIN A. HARRIS, M.D., D.D.S. Revised and Edited by FERDINAND J. S. GORGAS, A.M., M.D., D.D.S., Professor in the Dental Department of Maryland University. Eleventh Edition. 8vo, with 750 Illustrations, 31s. 6d.

A Manual of Dental Mechanics.
By OAKLEY COLES, L.D.S.R.C.S. Second Edition. Crown 8vo, with 140 Engravings, 7s. 6d.

By the same Author.

Deformities of the Mouth.
Third Edition. 8vo, with 83 Wood Engravings and 96 Drawings on Stone, 12s. 6d.

Notes on Dental Practice.
By HENRY C. QUINBY, L.D.S.R.C.S.I. 8vo, with 87 Engravings, 9s.

Elements of Dental Materia Medica and Therapeutics, with Pharmacopœia.
By JAMES STOCKEN, L.D.S.R.C.S., Pereira Prizeman for Materia Medica, and THOMAS GADDES, L.D.S. Eng. and Edin. Third Edition. Fcap. 8vo, 7s. 6d.

Dental Medicine :
A Manual of Dental Materia Medica and Therapeutics. By F. J. S. GORGAS, A.M., M.D., D.D.S., Editor of "Harris's Principles and Practice of Dentistry," Professor in the Dental Department of Maryland University. 8vo, 14s.

Eczema.
By McCALL ANDERSON, M.D., Professor of Clinical Medicine in the University of Glasgow. Third Edition. 8vo, with Engravings, 7s. 6d.

Diseases of the Skin :
With an Analysis of 8,000 Consecutive Cases and a Formulary. By L. D. BULKLEY, M.D., Physician for Skin Diseases at the New York Hospital. Crown 8vo, 6s. 6d.

By the same Author.

Acne : its Etiology, Pathology, and Treatment :
Based upon a Study of 1,500 Cases. 8vo, with Engravings, 10s.

On Certain Rare Diseases of the Skin.
By JONATHAN HUTCHINSON, F.R.S., Senior Surgeon to the London Hospital, and to the Hospital for Diseases of the Skin. 8vo, 10s. 6d.

Diseases of the Skin :
A Practical Treatise for the Use of Students and Practitioners. By J. N. HYDE, A.M., M.D., Professor of Skin and Venereal Diseases, Rush Medical College, Chicago. 8vo, with 66 Engravings, 17s.

Parasites :
A Treatise on the Entozoa of Man and Animals, including some Account of the Ectozoa. By T. SPENCER COBBOLD, M.D., F.R.S. 8vo, with 85 Engravings, 15s.

Atlas of Skin Diseases.
By TILBURY FOX, M.D., F.R.C.P. With 72 Coloured Plates. Royal 4to, half morocco, £6 6s.

Manual of Animal Vaccination,
preceded by Considerations on Vaccination in general. By E. WARLOMONT, M.D., Founder of the State Vaccine Institute of Belgium. Translated and edited by ARTHUR J. HARRIES, M.D. Crown 8vo, 4s. 6d.

Leprosy in British Guiana.
By JOHN D. HILLIS, F.R.C.S., M.R.I.A., Medical Superintendent of the Leper Asylum, British Guiana. Imp. 8vo, with 22 Lithographic Coloured Plates and Wood Engravings, £1 11s. 6d.

Certain Forms of Cancer,
With a New and Successful Mode of Treating it. By A. MARSDEN, Senior Surgeon to the Cancer Hospital. Second Edition. 8vo, with Coloured Plates, 8s. 6d.

Cancer of the Breast.
By THOMAS W. NUNN, F.R.C.S., Consulting Surgeon to the Middlesex Hospital. 4to, with 21 Coloured Plates, £2 2s.

On Cancer:
Its Allies, and other Tumours; with special reference to their Medical and Surgical Treatment. By F. A. PURCELL, M.D., M.C., Surgeon to the Cancer Hospital, Brompton. 8vo, with 21 Engravings, 10s. 6d.

Sarcoma and Carcinoma :
Their Pathology, Diagnosis, and Treatment. By HENRY T. BUTLIN, F.R.C.S., Assistant Surgeon to St. Bartholomew's Hospital. 8vo, with 4 Plates, 8s.

By the same Author.

Malignant Disease of the Larynx (Sarcoma and Carcinoma).
8vo, with 5 Engravings, 5s.

Clinical Notes on Cancer,
Its Etiology and Treatment ; with special reference to the Heredity-Fallacy, and to the Neurotic Origin of most Cases of Alveolar Carcinoma. By HERBERT L. SNOW, M.D. Lond., Surgeon to the Cancer Hospital, Brompton. Crown 8vo, 3s. 6d.

The Surgery of the Rectum.
By HENRY SMITH, Professor of Surgery in King's College, Surgeon to the Hospital. Fifth Edition. 8vo, 6s.

Lectures on the Surgical Disorders of the Urinary Organs.
By REGINALD HARRISON, F.R.C.S., Surgeon to the Liverpool Royal Infirmary. Second Edition, with 48 Engravings. 8vo, 12s. 6d.

By the same Author.

Lithotomy, Lithotrity, and the Early Detection of Stone in the Bladder ;
with a description of a New Method of Tapping the Bladder. 8vo, with Engravings, 2s. 6d.

Diseases of the Urinary Organs.
Clinical Lectures. By Sir HENRY THOMPSON, F.R.C.S., Emeritus Professor of Clinical Surgery in University College. Seventh (Students') Edition. 8vo, with 84 Engravings, 2s. 6d.

By the same Author.

Diseases of the Prostate :
Their Pathology and Treatment. Fifth (Students') Edition. 8vo, with numerous Engravings, 2s. 6d.

Also.

Surgery of the Urinary Organs.
Some Important points connected therewith. Lectures delivered in the R.C.S. 8vo, with 44 Engravings. Students' Edition, 2s. 6d.

Also.

Practical Lithotomy and Lithotrity; or, An Inquiry into the Best Modes of Removing Stone from the Bladder. Third Edition. 8vo, with 87 Engravings, 10s.

Also.

The Preventive Treatment of Calculous Disease, and the Use of Solvent Remedies. Second Edition Fcap. 8vo, 2s. 6d.

Also.

Tumours of the Bladder :
Their Nature, Symptoms, and Surgical Treatment. 8vo, with numerous Illustrations, 5s.

Also.

Stricture of the Urethra, and Urinary Fistulæ : their Pathology and Treatment. Fourth Edition. With 74 Engravings. 8vo, 6s.

Also.

The Suprapubic Operation of Opening the Bladder for the Stone and for Tumours. 8vo, with 13 Engravings, 3s. 6d.

Hydrocele :
Its several Varieties and their Treatment. By SAMUEL OSBORN, late Surgical Registrar to St. Thomas's Hospital. Fcap. 8vo, with Engravings, 3s.

By the same Author.

Diseases of the Testis.
Fcap. 8vo, with Engravings, 3s. 6d.

Diseases of the Rectum and Anus. By W. HARRISON CRIPPS, F.R.C.S., Assistant Surgeon to St. Bartholomew's Hospital, &c. 8vo, with 13 Lithographic Plates and numerous Wood Engravings, 12s. 6d.

Diseases of the Testis, Spermatic Cord, and Scrotum. By THOMAS B. CURLING, F.R.S., Consulting Surgeon to the London Hospital. Fourth Edition. 8vo, with Engravings, 16s.

Urinary and Renal Derangements and Calculous Disorders.
By LIONEL S. BEALE, F.R.C.P., F.R.S., Physician to King's College Hospital. 8vo, 5s.

Fistula, Hæmorrhoids, Painful Ulcer, Stricture, Prolapsus, and other Diseases of the Rectum :
Their Diagnosis and Treatment. By WILLIAM ALLINGHAM, Surgeon to St. Mark's Hospital for Fistula. Fourth Edition. 8vo, with Engravings, 10s. 6d.

Pathology of the Urine.
Including a Complete Guide to its Analysis. By J. L. W. THUDICHUM, M.D., F.R.C.P. Second Edition, rewritten and enlarged. 8vo, with Engravings, 15s.

Student's Primer on the Urine.
By J. TRAVIS WHITTAKER, M.D., Clinical Demonstrator at the Royal Infirmary, Glasgow. With 16 Plates etched on Copper. Post 8vo, 4s. 6d.

Syphilis and Pseudo-syphilis.
By ALFRED COOPER, F.R.C.S., Surgeon to the Lock Hospital, to St. Mark's and the West London Hospitals. 8vo, 10s. 6d.

Genito-Urinary Organs, including Syphilis :
A Practical Treatise on their Surgical Diseases, for Students and Practitioners. By W. H. VAN BUREN, M.D., and E. L. KEYES, M.D. Royal 8vo, with 140 Engravings, 21s.

Lectures on Syphilis.
By HENRY LEE, Consulting Surgeon to St. George's Hospital. 8vo, 10s.

Coulson on Diseases of the Bladder and Prostate Gland.
Sixth Edition. By WALTER J. COULSON, Surgeon to the Lock Hospital and to St. Peter's Hospital for Stone. 8vo, 16s.

On Rupture of the Urinary Bladder.
Based on the Records of more than 300 Cases of the Affection. By WALTER RIVINGTON, F.R.C.S., President of the Hunterian Society ; Surgeon to the London Hospital. 8vo, 5s. 6d.

The Medical Adviser in Life Assurance. By E. H. SIEVEKING, M.D., F.R.C.P. Second Edition. Crown 8vo, 6s.

A Medical Vocabulary :
An Explanation of all Terms and Phrases used in the various Departments of Medical Science and Practice, their Derivation, Meaning, Application, and Pronunciation. By R. G. MAYNE, M.D. LL.D. Fifth Edition. Fcap. 8vo, 10s. 6d.

A Dictionary of Medical Science :
Containing a concise Explanation of the various Subjects and Terms of Medicine, &c. By ROBLEY DUNGLISON, M.D., LL.D. New Edition. Royal 8vo, 28s.

Medical Education
And Practice in all parts of the World. By H. J. HARDWICKE, M.D., M.R.C.P. 8vo, 10s.

INDEX.

[Continued on the next page.

The following CATALOGUES issued by J. & A. CHURCHILL will be forwarded post free on application :—

A. *J. & A. Churchill's General List of about* 650 *works on Anatomy, Physiology, Hygiene, Midwifery, Materia Medica, Medicine, Surgery, Chemistry, Botany, &c., &c., with a complete Index to their Subjects, for easy reference.* N.B.—*This List includes* B, C, & D.

B. *Selection from J. & A. Churchill's General List, comprising all recent Works published by them on the Art and Science of Medicine.*

C. *J. & A. Churchill's Catalogue of Text Books specially arranged for Students.*

D. *A selected and descriptive List of J. & A. Churchill's Works on Chemistry, Materia Medica, Pharmacy, Botany, Photography, Zoology, the Microscope, and other branches of Science.*

E. *The Half-yearly List of New Works and New Editions published by J. & A. Churchill during the previous six months, together with particulars of the Periodicals issued from their House.*

[Sent in January and July of each year to every Medical Practitioner in the United Kingdom whose name and address can be ascertained. A large number are also sent to the United States of America, Continental Europe, India, and the Colonies.]

AMERICA.—*J. & A. Churchill being in constant communication with various publishing houses in Boston, New York, and Philadelphia, are able, notwithstanding the absence of international copyright, to conduct negotiations favourable to English Authors.*